AGRINDUS

AGRINDUS

Integration of AGRIculture and INDUStries

HAIM HALPERIN

Routledge
Taylor & Francis Group

LONDON AND NEW YORK

First published in 1963

Published in 2006 by
Routledge
2 Park Square, Milton Park, Abingdon, Oxfordshire OX14 4RN
711 Third Avenue, New York, NY 10017

First issued in paperback 2014

Routledge is an imprint of the Taylor and Francis Group, an informa business

British Library Cataloguing in Publication Data
A CIP catalogue record for this book
is available from the British Library

Agrindus
ISBN 0-415-38147-9 (volume)
ISBN 0-415-37652-1 (subset)
ISBN 0-415-28619-0 (set)

ISBN13: 978-1-138-87975-1 (pbk)
ISBN13: 978-0-415-38147-5 (hbk)

Routledge Library Editions: Economic History

AGRINDUS

AGRINDUS

Integration of Agriculture and Industries

by

HAIM HALPERIN

Professor of Agricultural Economics
The Hebrew University, Jerusalem

LONDON

ROUTLEDGE & KEGAN PAUL

First published in 1963
by Routledge & Kegan Paul Limited
Broadway House, 68–74 Carter Lane
London, E.C.4

Printed in Great Britain
by Lowe & Brydone (Printers) Ltd, London

© Haim Halperin 1963

Second impression 1964

CONTENTS

MAPS

PREFACE

THE purpose of this book is to present the problem of the village as a social, organizational, political, economic and technological category, which is in the throes of a crisis unprecedented in all its lengthy history. The stormy revolution we are witnessing at the beginning of the atomic age has brought the village to the brink of disintegration. A radical reassessment of values in many spheres of human activity is needed in our Era. For the village the issue is of far greater urgency especially in those countries which have been, or are rapidly becoming, industrialized. Basic and immediate changes are necessary.

When I was planning this book I found myself in a dilemma. Should I confine myself to a statement of the problem and its grave implications as I saw them, or should I construct a model as is consistent with the practice of modern science? Or should I adopt both methods? Most of the material for such a model—the physical, technical, organizational and other data—had been collected, the economic and statistical measuring instruments available. Israel offered an abundance of empirical material which could be used in such a model. I could use extensively various coefficients and quotients such as cost, labour, location coefficients; input-output analysis; linear programming; theories of growth; economies of scale; inter-industry and inter-regional relations; problems of resources, capital and demography.

But I was apprehensive of putting the stress on the technical aspect, which indeed interests many leading experts today. I preferred to deal with the phase preceding that of the model, which is more important than the model itself. No building can stand for long or even be erected, if the architect has erred in his static calculations. And if he has no aesthetic sense it will be ugly. If a structure is not animated by human activity it may be little more than a fortuitous combination of building materials. A model is hardly more than a stage in static calculation. It is significant only in so far as it leads to a more dynamic phase. It is completely valueless if its constructors make use solely of technical and physical instruments in erecting it and are not themselves inspired by an idea, and conscious that their model has an intellectual and moral mission.

I was also influenced in planning my work, by the consideration that present-day developments in regional co-operation in Israel have not yet reached the form which we are aiming at.

For these reasons I decided to devote this work to the conceptual model. If the idea is favourably received and the new proposed form is developed in practice (even if only in a few regions in Israel and perhaps in those other countries with whose leaders I have had the opportunity to discuss this problem), we shall have the empirical data to determine the magnitudes of the components of *agrindus*. Then we shall be able to proceed to economic analysis.

I am not aware, at the present stage, of any economic difficulties which might prove insurmountable. However, I do anticipate embarrassments of a psychological, traditional, cultural, social, political, and organizational character which might impede or even make our whole idea unworkable.

I have refrained from attempting to draft any formulae. I have adopted the two-way method of induction and deduction to explain an idea which is essentially socio-economic. For this purpose I have made use of living models with which I am familiar from direct contact and experience. These models already constitute centres in which agriculture and industry are integrated with an economic and social environment *sui generis*. I believe that in them we may already have adequate foundations from which we can develop variants of a method, and which we can vitalize and convert into forms capable of shaping existing processes of industrialization in agricultural countries.

Here I have tried to develop a method which, if accepted, can be elaborated in greater detail and translated into more practical terms, to correspond with the varying national, economic, social and other conditions obtaining in different countries and among different peoples. I envisage the introduction of many modifications—indeed such modifications are necessary and inevitable—in keeping with the ecological, social and cultural patterns of the world we live in.

I acknowledge with pleasure my debt to the many organizations, institutions and individuals who have shown so much courtesy and patience during the repeated calls I made upon them over a lengthy period, while the material for this book was being collected. The initial testing ground for these ideas was the *Mercaz Haklai* (Agricultural Centre) of the General Federation of Labour, which devoted a number of its meetings—in which many members participated—to hearing and discussing the basic principles expounded here. The

PREFACE

Mercaz even placed the issue on the 1959 agenda of a larger body—
the National Conference of the Agricultural Workers Organization.
More recently it has set up a special department to promote regional
co-operation and the establishment of joint regional economic
undertakings in keeping with my proposals. Two national confer-
ences of chairmen of regional councils in which senior officials of the
Ministry of Interior took active part were organized by that Ministry.
I have also explained my ideas to the Association of Engineers and
Architects (in Tel Aviv), the Jordan Valley and the Sha'ar Hanegev
Regional Councils.

I am particularly grateful to the Municipal Department of the
Ministry of Interior and to the Development Towns Division of the
Ministry of Labour for the abundance of restricted material which
was placed at my disposal, and also to the regional councils and their
Chairmen who gave every possible assistance.

Special thanks are due to Mr. P. Sapir, Israel Minister of Trade
and Industry, who showed a keen interest in the theme of this book,
took part in a number of discussions and promised me every assistance, both from himself and from the Ministry he heads, in the
development of co-operative industries in the various regions.

Among those who read the manuscripts, made valuable suggestions
and encouraged me in my task were H. Frumkin—Editor of the
Riv'on Hakalkala ('Economic Quarterly'); A. Zalel—Agronomist,
Israel Bank of Agriculture; Dr. Dan Yaron of the Hebrew University's Faculty of Agriculture; B. Kaplan—Agronomist, Director of
Settlement in the Negev; J. Gwirtz—Director of the Municipal
Division (who commented on the chapter on municipal problems);
Dr. P. Zussman of the Hebrew University's Faculty of Agriculture
(who made suggestions on the economic planning of *agrindus* in-
cluded in the final chapter); Mr. S. Rubinstein and Mr. J. Shiloh, who
are active in the affairs of Upper Galilee, read the section dealing
with that area. It is a pleasure to express my gratitude to Mrs. Sarah
Reichert-Gutgeld for her devoted secretarial help.

Finally I am deeply grateful to the Hebrew University for its
generous aid in making possible the publication of this book and
also to the publishers.

H. H.

I

INTRODUCTION

THE technological revolution, set in motion by the invention
of modern machinery in the early nineteenth century, gave a
tremendous impetus to the process of urbanization. In ancient
times, the few towns in existence had served as centres of culture,
commerce and administration. In the modern era, industrialization
lent a powerful stimulus to their development and expansion. The
populations of Europe and America, which in 1800 had been almost
entirely of an agrarian and rural character, began to gravitate at an
ever-increasing pace towards the new industrial areas. In the United
States, for example, in 1790 the population regarded as urban—by
definition then—constituted no more than 5·1 per cent of the total.
By 1850 this proportion had risen to 15·3 per cent, by 1940 to 56·5
per cent, by 1950 to 59 per cent; by modern criteria, according to
which any settlement with more than 2,500 inhabitants is regarded
as a town, it constituted no less than 64 per cent of the total.[1] In 1960
it stood at 88·5 per cent of the population.[2]

At the beginning of the nineteenth century England's urban population made up 10 per cent of the total number of inhabitants. By
1921 the figure had risen to 80 per cent,[3] and in 1961 was still 80
per cent.[4]

The United States and England, it is true, are extreme examples of
rapid urbanization, but in other Western countries, too, the prepon-
derance of the urban population has risen swiftly. Indeed, since the
Second World War, in almost all countries of a markedly agrarian
character on the Asian and African continents and in Latin America,
planned development—in itself a highly positive phenomenon—has
been accompanied by a process of industrialization. And the growth

[1] *Encyclopaedia Britannica.*
[2] *The Economist,* 10 Feb. 1962, p. 515, a diagram.
[3] *Encyclopaedia of Social Sciences.*
[4] *Britain*—official handbook (1962 edition).

1

of the towns has always been a concomitant of industrial expansion.

The village is not merely an economic entity. It represents a way of life. And despite all the patent marks of class differentiation in the village, in both agrarian and industrialized countries, rural society has always conjured up associations of a special character.

In recent years sociologists have argued in favour of the existence of the *urban community* and its development. It may well be that although towns are often characterised by the diversity of occupations, income groups, standards of living and culture among their inhabitants, and by their national, communal, religious and lingual heterogeneity, they nevertheless possess a certain common denominator which justifies the use of the term 'urban community'. But at the same time two radically opposed trends can already be discerned, one towards urbanism and the other towards ruralism.

These trends have been instructively summarized in the following terms:[1]

'There are two community sociologies, rural and urban . . . The field of rural sociology is rural society and rural living while that of urban sociology concerns society and living in towns and cities. . . . Urbanism as a way of life is primarily, although not entirely, associated with living in towns and cities. It differs from ruralism mainly in ways of work, habits of thought, and with respect to traditional controls. The tempo of life and work is faster under urbanism. Urbanism is the more dynamic of the two while ruralism is the more self-isolating . . . in the city, urbanism has been described as superficial, anonymous, and transient . . . Different elements of urbanism may be present in a different degree, depending on the relative mixture of ruralism and urbanism in particular areas . . . Even though the rural way of life changes to become more urban-oriented, *rural* as a way of work and life remains. . . .'

In many languages innumerable words have evolved from a term describing a single trait or characteristic to define, *pars pro toto*, an entire concept. In all languages, however, the terms 'village' and 'town' are so plain and unambiguous that there is no need for any further qualification.

But it is not with the words but with their meaning or content that we must deal in this context.

We are discussing a concept closely related to economics, technology, sociology and culture—and in certain countries even to national security.

Needless to say not even the most radical urbanist is totally

[1] Nels Anderson, *'The Urban Community', a World Perspective*, Routledge & Kegan Paul, London, 1960, pp. 21-3.

opposed to the village and to rural values, and by the same token no ruralist entirely opposes the town and its way of life.

Towns are increasing in number and size, while the villages are shrinking steadily. Before the Age of Steam it is doubtful, indeed, whether there was a single city in the world with more than one million inhabitants. Today there are not a few multi-million cities, where people reach their homes by lifts and make their way either through or under the streets in automobiles or railways. Children grow up without playgrounds, without trees or lawns and without knowing the joy of playing with a domestic animal. Their toys are made from springs, motors and electric batteries. Everything they handle is mechanized or is modelled on machines. What does the future hold for a child born on the 57th floor of a building in one of these vast urban conglomerates, whose very childhood is 'mechanized' and who, when he grows up, may be employed in one of the many branches of electronics or, worse still, minding some automatized machine? He will be the New Man! But what sort of a man will he be? Soon, perhaps sooner than we think, we shall have the privilege of making his acquaintance. The Towers of Babel which have already emerged and which may yet be erected intensify the inability of human beings to communicate. They will atomize human society. *Homo homini alienus* may describe a far crueller relation than *homo homini lupus*. Anxiety for the future and welfare of their children gives the town-dwellers no rest. Indeed, many parents, repelled by life pent-up in the cities, in the towering skyscrapers, dream of a small cottage in some quiet, grassy suburb.

But the decline of the village continues swiftly. In the United States the rural population dropped by 12 million between 1910 and 1957, while the proportion of the rural population declined by 34·9 per cent to 11·9 per cent of the total.[1] In England, to cite another example, the rural labour force in the period from 1881 to 1951 dropped from 13 per cent to 5 per cent,[2] and by 1955 it had declined further to 4·5 per cent. This, of course, did not cause any reduction in agricultural output. The contrary was true. Production rose to meet the growing consumption of a rapidly increasing population. There is, of course, no reason to fear reduced farm output, for the world will always need staple commodities. When the ages of steam and electricity have passed, perhaps in the course of, or after, the atomic age, when modern technology has been harnessed to meet the needs of human existence, it will prove feasible to produce organic foods without soil, to store the sun's energy in factories without using vegetable matter and to produce high-quality protein without

[1] U.S. Dept. of Agriculture: *Agricultural Statistics, 1957*, Washington, 1958.　　[2] Source: *The World's Working Population*.

3

passing vegetable matter through the stomachs of animals. At some future phase of the technological revolution cultivation of the soil may prove unnecessary, and derivatives of the Latin word *ager* will fall out of use in living language.

In the present essay, however, we do not propose to indulge in any Utopian fantasies, nor even to predict the shape of things to come in the nearer future. We are concerned here with the phase, upon the threshold of which we now stand, in this second half of the twentieth century. Ours is a bewildered generation. We have scaled the heights but it seems that the ground is slipping from beneath our feet. We still stand in need of the *ager*. Without it there can be no cultivation. In our age agriculture has assumed vast proportions, but the village is beginning to vanish from the face of the earth. The technological revolution, industrialization and high levels of productivity are attracting more and more people away from the orbit of agriculture and thereby from the rural community. Industry and agriculture, town and village, and technology even more, are the ripe fruit of the human spirit. Man created them, they exist to serve Man. But agriculture and the village are in the throes of crisis, for industry has become dominant. Are we so certain that no crisis is in store for industry and that we have not gone beyond the bounds of prudence in concentrating millions of human beings in single cities, in multi-storeyed structures, in vast industrial plants? Are we sure that they are safe from any cataclysm of nature or economics? And what will the city look like when the population of the world multiplies from three to six, and later to over nine billion? Picture a city of 25 or even 50 million. We may assume that our town-planners will find some way to organize traffic, housing and life generally in such vast urban centres. At the moment we lose at least one hour daily in traffic jams as we are on our way to or from our work. So far no satisfactory solution has been found; it must be found when our cities double and triple their populations. We have indicated that this is a problem we must leave to the architects and town-planners. But it is less easy to thrust our concern about the urban community on to the sociologists. The latter, up to the present at least, have not developed society; they have merely described it, analysed its characteristics and traced its trends. It is difficult, if not completely impossible, for any man in our own age to imagine the future city with its teeming millions. Can it possibly be an integrated unit? Are not trends towards disintegration more likely in this atomic age? What will be the fate of the children of these vast cities, who no longer suffer from the sentimental inhibitions of their parents who look back to a more idyllic period? What will our great-grandchildren look like? Even the small and medium-sized towns have already atomized society. I live in the largest city of a very small country. When I settled in it, not many decades back, it had no more than 15,000 inhabitants. Within a short time of my arrival most of the citizens knew me—and I them—personally. But things changed as the town expanded, especially in the thirties. For twenty years until the middle fifties I lived in a quiet street, containing thirty-six buildings, not one of which was more than three storeys high. My official and other interests, like those of most of my neighbours, were outside this residential area. I had little contact with the other tenants living in the same street, except for the grocer, the shoe-maker and the laundry man. Over a period of twenty years I had never learned the names of my neighbours living in the buildings opposite and flanking my own. Seven years ago I moved to a new building. In the same street I had a neighbour—no relative—with the same name. We became acquainted as a result of the postman's errors. One day when I was walking down the stair-case with my wife I saw a black-bordered funeral notice, bearing my own name. My neighbour was dead. But however much she tried my wife could not remember what our deceased neighbour had looked like though we had lived close to one another for six years and even bore the same name. We had resided together but had not lived together. Even death had not left its mark.

Our natural weakness for gossip is not limited by geography. It extends to our places of employment and to the political party we belong to. But it is totally different from the type of gossip about what goes on in our neighbours' kitchens. In a village with, say, a hundred families, everybody knows everyone else. If there is illness or misfortune in one of the families, then all are concerned. The ninety-nine families not directly affected are moved by a genuine sympathy for the unfortunate one. It is the same when there is any occasion for rejoicing. All rejoice together with the happy family. The heart beats differently in such a small settlement.

Not long ago I attended the funeral of a friend—a member of a *kibbutz*. I was deeply impressed by the obvious grief of the 442 men, women and children of the settlement. Moreover hundreds of settlers from neighbouring *moshavim* and *kibbutzim*—not to speak of friends from a greater distance—had come to pay their last respects. Hundreds attend a funeral in the country; in the big city, however, it is a small affair, the mourners comprising hardly more than the imme-diate relatives. Man's loneliness in the city, in both life and death, is a terrible thing.

What is the impact of the city upon children? The school, or their educational environment, whether good or bad, is not enough.

But why should we give up rural society entirely? Are we indeed capable of developing a better, more integrated society; one that is

5

more robust—not only physically but also morally—more patriotic and more upright? Are there not more neighbourly relations in a village, and more effective mutual aid, sympathy and concern among the inhabitants?

Yet the village is shrinking steadily. The International Labour Office has made a world survey of the reasons for the desertion of the village.

'. . . The movement out of agriculture is a general phenomenon,' the published Report points out. 'Labour is leaving the land in all countries. Everywhere the share of agriculture in the total labour force is declining. . . . An absolute decline in the size of the agricultural labour force has been proceeding in the advanced countries for varying periods—for a century in France and the United States, from the 1930's in Australia, Canada, Denmark, Germany, New Zealand, and Norway, and since 1947 in the Netherlands. But in the last ten or twenty years the decline has greatly accelerated.'

'If people are asked to explain why they have left agriculture, they usually give a variety of reasons. Several direct investigations have been made in different countries in recent years. . . . The reasons actually given include better pay, shorter hours, better educational and transport facilities, mechanization, the impossibility of supporting a family on a very small holding, the difficulty of rising in the social scale or of gaining access of ownership, or simply the shortage of rural housing.

To list all these various reasons as separate factors determining movement in different conditions tends to obscure the importance of the two main factors which are common to all countries and within which it is possible to include most of these reasons for migration. These two main factors are, first, the level of relative incomes and, secondly, the opportunities for non-agricultural employment.' [1]

Is there no way of halting this trend away from the land, by removing, or at least reducing, the causes? Is it in the interest of society, the State or the world at large that the village should disappear? The village is menaced by the steady encroachment of urbanization. In Israel, hundreds of new villages, based upon agriculture, have been established in the past decade. In this period twenty-five new towns have been laid out, while the prospect of founding another dozen in the Negev—the arid, southern territory of Israel—is under consideration. These forms of economic development may be an unavoidable necessity. But let us try to think out some real solution,

[1] International Labour Office. *Why Labour Leaves the Land.* A comparative Study of the Movement of Labour out of Agriculture, Geneva, 1960, pp. 11–14.

if not to preserve the village as we have known it up to the present, at least to achieve some compromise to serve ourselves and our children. Future generations will have to find their own path towards the good life, in accordance with their aspirations and the conditions obtaining in their time.

Agriculture can be combined with industry without undermining that age-old social asset—the village. We can improve and even reform the village and bring it into line with changing conditions.

The rural family, which today at least is larger than the family in the city, has a surplus of labour owing to the advance of agrotechnology and higher levels of productivity. This surplus must find an outlet in other, non-agricultural, branches of the economy. But why should the sons and daughters of the land be compelled to leave their homes, their parents and relatives? Why not develop non-agricultural sources of employment and livelihood in the immediate neighbourhood, so that when they have finished their eight hours of work (or seven or less as the case may be in the future), they can return home and pool their earnings with those brought in by other members of the family employed in farming? Those not employed in agriculture will enjoy all the benefits of village life, while if they want the amenities of the city they will find them near at hand. The industrial centre will be set within a cluster of villages and equipped with a cinema, a theatre, cafés, cars and garages. They can listen to the radio or watch television in their own homes. A rural family which derives its income not only from agriculture but also from industry, commerce and clerical employment, will enjoy a higher and more secure standard of living. The flight from the village will be checked and the deep-rooted and closely-knit rural society, which, at the moment, has a more definite image than the amorphous urban conglomerate, will be preserved. Up to this day we know of no city which has had rural characteristics. But the converse, a village possessing elements of the town, is feasible. Moreover, it is far more desirable to establish an industrial centre in a rural setting. Within a group of 25, 30 or even 40 villages, let us say, an industrial centre would be developed possessing all those attributes of a city which the villagers find so attractive. The urge towards mobility, so deeply rooted within us, would also be satisfied.

The giant city, the vast economic corporations, the ramified institutions and the irresistible urge towards centralization all wear down and ultimately destroy the smaller economic units and with them the smaller social units. The rapid advance in economic life and technology makes for increasing centralization. But the human spirit cannot master these vast creations; it needs decentralization. The basic cells of society are, in the meantime, steadily becoming attenuated; the general fabric can only live if its cells are constantly renewed. The same is true of the human fabric. Increasing centralization has not only weakened the family unit; it has weakened the ties holding the units together. Perhaps the units as they are constructed today no longer fulfil their function. Society may need new, stronger units, more accommodating to the human spirit. Then, we may hope, education will also prove more capable of achieving its objectives.

Developments during the past hundred years have tended towards a degree of centralization which we regard as excessive. We must try to create intermediate forms, socio-economic combinations capable of preserving the integrating power of the basic cell. The cells will combine to form the core of the larger organism. How are we to develop these new cells or rather how are we to enable the existing cells to renew themselves, to develop a form that will protect them under the new conditions emerging?

In Soviet Russia, at one time, the idea of developing an *agrogorod* —an agricultural town—was broached. What was envisaged, it seems, was a vast *kolkhoz*. The notion was not taken any further, however, and it is not even clear what form its authors intended it to take.

We have little real knowledge of developments in People's China. The plans for the development of people's communes provide for a combination of agriculture, industry and security. The prospect of establishing urban communes has also been considered. But the dimensions of what is being undertaken are so immense and the momentum so great, that a long time must go by before we shall be able to pass any judgements.

During the forties of this century another project for integrating agriculture and industry—*urbanisme rurale* or *rurban*[1]—was put forward. It too, however, remained undeveloped. We know of no attempt to incorporate its principles. The establishment of factories in rural areas as in Switzerland, in some parts of the United States, and in other countries does not, it seems to us, offer any answer to the question we have posed. We suggest something new: the development of areas comprising a score or more of villages, ranging around an industrial centre, in which, in the course of time, other non-agricultural enterprises will be established. Thereby the gap between the urban and the rural standard of living will be narrowed, agriculture will be preserved, and rural society will continue to exist though it will assume different forms.

What name shall we call this child? If a name is so important it will not be difficult to find one. We are discussing the integration of agriculture and industry. Etymologically the dominant roots provide

[1] Wilhelm Abel: *Agrarpolitik*, Göttingen, 1958, p. 244.

8

a clue. Let us call it AGRINDUS. It may seem idle to look for a name before the child is born, but in the present instance the name indicates a trend and embodies an idea. Any other name would serve, if it were more expressive. What is important here is the substance, the content.

In embryo *agrindus* already exists in Israel. It must be given proper pre-natal care; it must be nourished; precautions must be taken against miscarriage, to ensure that it is born healthy and that it grows up to be a joy to its parents and to society.

II

THE THEME

A RECENT development in Israel, which in its new co-operative economic form has just emerged from its infancy in agricultural areas, can serve as a sound foundation for the growth of the *agrindus* idea. If it is allowed to develop on the right lines it may serve as a model for other countries, especially those of a preponderantly agrarian character, which are eager to develop their industry and to share in the process of industrialization which is sweeping the world. If, as may well be, industrialization assumes excessive proportions, the form proposed here will mitigate and perhaps even obviate its more painful features.

Highly developed industrial countries will not be interested in our proposals, though it is not entirely impossible that in the settlement of wastelands or sparsely populated areas they would be prepared to contemplate new methods. The Tennessee Valley, for example, might have been settled upon the lines proposed here. Similar opportunities will certainly recur in the United States and Canada, and perhaps even in England, as well as in other countries such as those of Eastern Europe and Latin America. It may be expected that the birth pangs will be difficult in both agrarian and semi-industrialized countries, though here the prospects for the method we propose are better.

As far back as the twenties we, in Israel, embarked upon the integration of agriculture, industry, building, public works and security within a single collective unit, the dimensions of which were comparatively modest. Dozens of *kibbutzim* still continue to operate workshops and industrial enterprises in mainly agricultural settlements. In our own time, however, factories must be planned on a larger scale than is possible in a single settlement of several hundred members. Collaboration between a larger number of *kibbutzim* and *moshavim* can make possible the establishment and development of such larger enterprises—not only factories, but technical, economic

10

and cultural services, the operation of large machines, cold storage plants and slaughter and packing houses, for example—which are beyond the capacity of any single village, whatever its size. In essence what we propose is comprehensive regional co-operation which will allow the development of plants not much smaller than those in the cities.

We do not wish to pre-determine any form of ownership for these plants in the industrial centres of these rural areas. In Israel, where there are many co-operative villages, it will not be difficult to establish industrial centres co-operatively run by all or some of the villages in any district (in the event of there being villages not wishing to join). Every village, it need hardly be said, will be at liberty to decide freely whether and when to join the co-operative centre. If it does not join at the start it will probably do so when it sees the benefits that can accrue. Important beginnings in this direction have already been made in Israel, as will be shown in later chapters. Partnerships between villages and private capital in the operation of factories and services can be formed in these centres. Israel offers a number of examples of such co-operation. There can also be enterprises that are privately owned, or indeed under any combination of ownership, according to the feelings and wishes of the local environment.

In the chapters which follow we shall describe developments in this direction in Israel. We shall give an outline of what has been achieved and then propound the changes suggested by us. Briefly we shall portray *comme il est* and then go on to *comme il faut*. Finally we shall suggest an outline which is amenable to modification and adjustment to local conditions and preferences.

In Israel, it is true, planners and others are not looking in the direction we suggest. In spite of this we do not regard our plan as an intellectual exercise. Once we have traced the various phases in the resettlement of Israel from the closing decades of the nineteenth century to the present, we shall see that not much remains to be done to bring existing realities into line with what we propose. To us it seems that we have here an almost organic process from the beginning of the modern return to Zion by way of rural and urban settlement, and various intermediate phases; for what we envisage is not a short-cut but a planned and phased advance. The social forms evolved and the economic operations undertaken without prior planning or forethought come to form an organic pattern. In the present stage with a little steering we can arrive at a completely rational system for our age.

In 1844, in Rochdale, twenty-eight labourers laid the foundations of consumers' co-operation. They had drafted no programme and were completely ignorant of the significance of their step. In 1910 twelve new immigrants, ten young men and two young women, banded together to form a working group, which, to use their own down-to-earth phrase, 'would have one pocket and one plate'. They worked as hired labourers on farms, and after a dispute with the manager of one estate where they were employed, the Director of the Palestine Office (the owners of the estate) allowed them to undertake cultivation on their own account and responsibility. This was the modest beginning of the first *kvutza* (collective settlement) in Israel. The founders of Degania A (as they named their tiny village) could hardly have foreseen then that within fifty years the movement they had begun would count 228 settlements with 85,000 inhabitants!

It was only in the forties that we broached the idea that the *kibbutzim* combine to establish regional co-operative economic enterprises. The *kibbutzim* themselves were hesitant; some preferred to develop their own workshops and factories. Twenty years later, however, a movement to establish regional plants—this time emanating from within—was in full swing. What is necessary is a further advance, which is what we now propose. The need will be understood better now than it was twenty years ago, as we have learnt much from experience.

We are aware of the widespread interest in the reciprocal influences of agriculture and industry. We know of the surveys that have been made on the impact of an industrial plant established in an agricultural district, as in certain counties of Louisiana and Utah. We have spoken at length on our subject but we have not yet come to the essential point which we propose to develop in this study. Some time ago the Israel Ministry of Interior sponsored two conferences of Chairmen of Regional Councils. I was asked to address the gatherings on the *agrindus* idea. My thesis was welcomed enthusiastically; the proposals I had to make were described as visionary, but the participants indicated their willingness to help translate them into reality. However, they are still far from the methods we urge. In many areas not even the first step has been taken in this direction. In certain cases no plans have even been prepared. Where regional projects have been established they are purely adventitious, the fruit of individual initiative or external circumstances. The service undertakings are the more prominent. The projects, generally, are owned by limited partnerships of a small number of settlements in the area, and not by the entire district. It is hardly necessary to point out that there is no co-ordination between the various districts. An important enterprise combining agricultural services and processing industries being developed in Samaria (see Chapter VIII) jointly by 38 *kibbutzim*[1] situated between Haifa and Tel Aviv offers an instructive

[1] A *kibbutz* is a group of individuals who have united voluntarily for the

12

example. Ownership of this plant goes beyond regional boundaries. It includes settlements in six districts, but no *moshavim*,[1] which for some reason have been unwilling to join. But what is most interesting is that such a large and important project can operate only on the economic level. In view of its comprehensive character, its distance from many of its affiliated settlements in the north and the south, it can make no sociological contribution—and it is this aspect that is the principal characteristic, the main pillar, of the socio-economic structure which we propose. The social aspect cannot be divorced from the economic, nor can these be isolated from the educational, cultural, political and other aspects.

In the meantime, however, new developments have taken and are taking place whose course it will be hard to alter. We are under no illusion about the difficulty of our task, and it is the most difficult aspect that we shall tackle first. Nevertheless, we shall propose models for areas which for our purposes may be regarded as *tabulae rasa*. If, in the course of time, the advantages of this method are proved, it

[1] The *moshav ovdim* is an organized society with a common aim and goal. It is based upon the individual labour of the farmers with the aid of the members of their families. Their way of life is governed by prescribed social principles, among the most important of which are: national ownership of the soil, self-labour, mutual aid and co-operative marketing of produce and purchase of supplies and equipment.

The *moshav shitufi* is a type of village where consumption is individual but agricultural production is on a collective basis (the degree of which may vary from village to village). In some of these *moshavim shitufiim* all branches of farming are conducted collectively, in others this is the case only in the major agricultural branches. *Moshava*—an ordinary village with individual farms on privately owned land, with or without hired labour, where membership in Co-operative Societies is not obligatory.

For more detailed description see: H. Halperin, *Changing Patterns in Israel Agriculture*, Chapter XI, pp. 201–50 (Rural Sociology in Israel). Routledge & Kegan Paul, London, 1957.

purpose of establishing in Palestine a homeland for the Jewish people based upon socialist principles. To achieve this goal, they have created a mainly agricultural, economic and social unit embodying the principles of complete equality, mutual responsibility and self-labour. The ownership of private property has been abandoned and all production and consumption organized on a collective basis. The *kibbutz* requires every member to make his contribution according to his capacity and allows him to receive from the common fund in accordance with his needs. A *kibbutz*, therefore, is a collective village in which all property is owned in common by its members, who pool their resources and provide all necessities out of a common fund. In addition to the main occupation—agriculture—there is often some industrial enterprise.

can be adapted to meet constantly changing conditions. In a country combating the wasteland, moving mountains and diverting the course of rivers, in a generation which has witnessed the liberation of peoples, the immense progress registered in the physical world, the invention of satellites and rockets, and the conquest of space, some means must surely be found of filling the void in the economic and social spheres, in industry and agriculture, in town and in country.

We are sanguine enough to hope that once the project we have outlined takes final shape in Israel, where favourable conditions have almost matured, it will also be found suitable for other countries, if not immediately then in the course of time.

Let us examine the proposal carefully and perhaps we shall decide upon some course of action.

In what follows we shall outline the various phases, which while mutually dependent—certainly in the manner in which they crystallized—in fact now constitute milestones on the road to *agrindus*.

The *g'dud ha'avoda*—the labour corps, founded in Palestine in the early twenties—which we shall have more to say about in Chapter IV—seems to us to be the progenitor of the idea. The industrial complex developed by the Histadrut, the General Federation of Labour, and particularly the factories in the workers' settlements, were undoubtedly a natural development of the Labour Corps. The idea was carried a stage further when agricultural settlements were organized on a regional basis, first for municipal purposes, then to lend the new municipal body more substance, to institute and maintain services which the individual villages could not undertake separately. Then began the processing of local raw materials, the construction of large storage facilities (including refrigeration and freezing plants), poultry feed stores, cotton gins, lucerne meal mills, curing and packing houses, garages, metal and carpentry workshops, and finally the development of factories (for plywood, canned goods, etc.) whose products are sold on the domestic market or exported. Most of these undertakings were organized within a single district, but in certain cases the regional organization took in several districts.

These regional bodies provide the natural foundation upon which to base the *agrindus* idea. We shall outline and examine each phase. Then we shall go on to describe a number of districts in the northern, southern and central areas of Israel—which have already been organized on these lines. Later we shall proceed to draft an outline of *agrindus* as we conceive it.

What is the purpose and what are the main principles of *agrindus*?

(1) The purpose of *agrindus* is to secure co-operation between the largest possible number of neighbouring villages for the maintenance of agricultural services, the processing, storage, grading, packing,

14

transport, marketing and financing of farm produce, and the establishment of factories and workshops to meet agricultural and other requirements. This co-operation must be extended to include cultural and sporting facilities, educational and health institutions and the like. At the highest level, co-operation in farm production is, of course, also possible.

(2) Essentially *agrindus* implies the combination of agriculture and industry.

(3) *Agrindus* is a regional rural unit, combining the largest possible number of villages of various types on the basis of regional co-operation. A town is at the centre of such a region.

(4) The limits of municipal jurisdiction should coincide with those of the *agrindus*, even if the delimitation of municipal borders has to be altered for this purpose, since the socio-economic integrity of the area is of paramount importance. It is therefore highly desirable to have municipal boundaries demarcated or altered after the borders of the *agrindus* have been finally determined.

(5) Every village—*moshava, moshav, moshav shitufi* or *kibbutz*—shall confine its economic activities within its own area mainly to farming and the operation of ordinary services. Non-agricultural enterprises which already exist need not be liquidated provided they do not make excessive demands on locally available manpower. In future, however, such undertakings should not be established within the confines of the villages.

(6) Every village can continue to foster its own way of life as a *kibbutz, moshav, moshav shitufi* or *moshava* as the case may be. At the same time it will co-operate with all other settlements in the region in the economic and municipal spheres, and, if and in so far as it wishes, in the cultural and social spheres as well.

(7) In order to raise the productivity of labour and to reduce costs large-scale, regional co-operative undertakings will be developed outside the normal branches of farming, such as cultivation of fields, gardens and orchards, the breeding of cattle and poultry, apiculture, etc. The individual farmer need no longer prepare his own poultry and cattle feeds, ripen his own bananas and dates, put his apples and potatoes into cold storage, freeze his chickens, gin his own cotton, grind his own corn. It is not economical for a farmer to buy a heavy tractor which he cannot use all the year round. All these are operations which can be undertaken on his behalf, more efficiently and at lower cost, by regional plants in which he holds a share.

(8) Nor is it economical for a small village to build its own little school. The regional school will be more efficient and cheaper. There is no need for the village to maintain its community centre, for the facilities the regional amphitheatre, in which it is a partner, can offer

15

by far exceed anything a single village can aspire to, and here too the cost will be lower. The amphitheatre will be available for performances by national theatres, operatic companies, orchestras and choirs, and also by artists coming from abroad.

All enterprises of the type set out in paragraphs 7 and 8 will be located in the regional town, which is an essential feature of the *agrindus*.

(9) As already stated, every village will retain its specific agricultural character. Non-agricultural activities, both those which are related to farming and those which have little or no connection with farming, such as, for example, metal and carpentry shops, plastics, spinning, weaving and sewing factories, will be concentrated in the regional town. Villagers (members of *kibbutzim* or *moshavim*, or their sons and daughters) can work in these factories and workshops, in the marketing agencies and in the regional bank. They can be teachers, doctors, clerks, scientists on the staff of an experimental station, labourers at the waterworks and drivers or mechanics at the local garage; they can seek employment in any local service. Their earnings will suplement those of their families or of the co-operative of which they may be members, in exactly the same way as if those earnings had come from the family or collective farm. They and their children can continue to live in their village, with all the rights and duties entailed like all other settlers in that village or members of that collective settlement. The important advantage of this system will be that the regular wages of the member of the *kibbutz* or *moshav* or of their sons and daughters will go into the common pool together with the income derived from farming.

(10) Organized transport will enable residents of the villages to go to and from their places of employment in the regional town in less time than is usually the case in a large or even medium-sized town. On the basis of experience in a number of regions the journey to or from work will take no more than 15 minutes in a region comprising 30 settlements.

(11) The regional town—the town of the *agrindus*—must be in complete harmony with its environment as, in essence, it constitutes the expression of regional co-operation between the villages, which, too, are mainly of a co-operative character.

(12) We have already indicated that the regional town, which, as far as possible, in order to facilitate internal transport, will be located in the geographical centre of the district, will furnish the base for distributing heavy agricultural and other machinery, such as tractors, bulldozers, shoveldozers, graders, rooters and combined harvesters, and will provide for the area, garages and spare-parts stores, trucks, cold storage plants, packing and grading houses,

16

slaughterhouses, freezing plants, cotton gins, feed-mixing plants, curing houses and every other undertaking required for the processing of farm produce.

(13) The operations enumerated in the previous paragraph can be classified as services. In the regional town there will be plants carrying the processing of farm produce a stage further. These include flour-mills, bakeries, spinning and weaving mills, dyeing plants, tanneries, garment and shoe factories, oil presses and plants for oleaginous products, canneries, pharmaceutical and perfume factories, etc. Bigger plants can be established in partnership with other regions. These would include sugar refineries, timber, metallurgical and paper industries and indeed factories of any and every kind.

(14) The services and industrial plants enumerated in paragraphs 12 and 13 combine to form a ramified complex, lending the town of *agrindus* its essential character—that of an industrial centre in an agricultural context.

(15) Marketing, supply and transport facilities will be concentrated in the town.

(16) Regional administrative offices, including postal, telephone and telegraphic services, personnel for instruction and inspection in agriculture, industry, commerce, cultural activity, public health, etc., will also be based upon the regional town. The regional clinic and hospital, as well as other public health facilities, will be there. Banks and national economic institutions will establish local branches. The General Federation of Labour, the Agricultural Organization, the settlement movements, the political parties, newspapers and publishing houses will all have branches in the *agrindus* town.

(17) Municipal organization will be adapted to meet the special needs of *agrindus*. The town, together with all constituent villages, will make up a single municipal district, headed by a Council of representatives of all settlements. The Regional Executive will be elected by the Council. The latter, in keeping the laws governing municipal administration, will come under the supervision of the Ministry for Interior. The seat of the regional administration will, of course, also be in the regional town. The town, it must be stressed, will be represented on the Regional Council; under no circumstances will it constitute a separate local authority. Indeed the Ministry of Interior must refrain from granting the status of 'local council' to any settlement within such a region, and will endeavour to correct errors in municipal organization made before the *agrindus* idea is finally realized.

Experience has shown that a certain conflict of interest has developed between the villages in the area and the regional town. The latter was originally planned to serve the region in which it was

17

located; pressure of immigration, however, and the gravitation of villagers stimulated excessive expansion beyond the needs of the region, and raised a demand for employment not necessarily serving the rural hinterland. This has always been the case when the town has insisted upon secession from the regional council and independent representation, divorced from the neighbouring villages. Almost all of the development towns went through this process, beginning with their devoted adoption by and period of guardianship under the regional council and concluding with an insistence on the part of the residents upon their right to municipal autonomy.

Ministry of Interior officials are contemplating a solution by way of establishing a district council, i.e. the erection of another storey upon the existing structure of local government. The district council would embrace the regional council, the smaller local councils and the township council. This 'roof' council would plan for and serve the villages and the townships of an entire district. The district council would deal with all matters beyond the capacity of the regional and the local councils such as hospitalization, drainage, regional sewerage, water authority, physical and economic planning.

The district council would replace the district officer, who would become the chairman of the former. The council would also assume functions at present carried out by district offices of the Ministries of Health, Agriculture, Social Welfare, Education and Culture, and might even take the place of the Towns Associations.

(18) The Regional Councils in Israel have initiated and sponsored various economic projects in their areas. To meet legal requirements different companies and associations the directors of which are the members of the Regional Executives have been established. In addition, the Councils themselves have also acquired shares in such enterprises. This is the result of peculiar circumstances. Economic operations on a regional basis might never have been instituted had it not been for the initiative of the Regional Councils. This state of affairs can continue in the future. But as the regions expand as proposed in a later section (Chapter VII—Municipal Organization), the Councils should confine themselves not only legally but in fact to their municipal administrative functions as these are bound to increase.

(19) Who will be the inhabitants of *agrindus*? On the face of it *agrindus*-town will be of a purely functional character, as all operations will be undertaken by the settlers in the villages, who are the members of *agrindus*. This principle is important from both the social and the personal standpoints. In the same way as the town is an integral feature in the planning of an agro-industrial district, the inhabitants must fit in with the special economic and social con-

ditions in that district. It must be taken into account that in every rural settlement there are people, especially among the younger generation, who prefer non-agricultural occupations. Moreover, constantly rising labour productivity is reducing the manpower required for farming operations in any village. The *agrindus*, combining agriculture with various other occupations and creating new outlets for the rural inhabitants in the district, will halt the flight from the villages. The sons and daughters employed in the town will return to their homes and families in their villages when the day's work is done.

(20) There will, of course, always be people in *agrindus*-town. The nature of their employment and their duties will require many of them to be there. Some factories will work in three shifts. Other plants will keep workers on at night to clean-up or to do such tasks as running repairs. There are services that must be maintained round the clock, such as the water supply, power, sanitation, public health, postal and transport services and refrigeration. But *agrindus*-town need not have permanent residents.

(21) Outside factors or, say, external pressures, such as, for example, mass immigration and the need to absorb the newcomers as rapidly as possible, the exigencies of wider population distribution or the presence in the district of many people who are reluctant to belong to any village and prefer to work and reside in *agrindus*-town, may require modification of the principles set forth in the previous paragraph. Wage earners may prefer to live in the town. 'Development towns' have sprung up in agricultural districts in Israel as a result of outside pressures and willy nilly we must adjust ourselves.

(22) As far as possible enterprises in the district town must be *owned jointly by their sponsors and employees.* Every undertaking will be based upon a co-operative society or company (a society is preferable), the members of which are villagers in that district. The co-operative society which constitutes the legal body of the economic enterprise in the region can include as share-holding members the Regional Council, a national co-operative society (whose collaboration the enterprise may be interested in), e.g. Tnuva (Hebrew: produce), which is a national co-operative for the marketing of agricultural produce, or Hamashbir Hamercazi (Hebrew: central supplier), a wholesale co-operative for supply and consumption, or a national construction and contracting co-operative. It would be highly desirable for all economic and other enterprises, organized in the form of co-operative societies, or more precisely, all societies representing the enterprises, to be members of a single, regional roof co-operative. There would also be room in the latter for the Regional Council and the Executives of the National Co-operatives affiliated

to the movement in which most of the villages in the district are also members, such as the Hevrat Ha'ovdim of the General Federation of Labour or Nir Shitufi of the Agricultural Workers Organization, in which the primary societies are organized. The two central co-operative societies (Hevrat Ha'ovdim and Nir Shitufi) which are the mother-societies of the primary societies will hold the right of veto, in keeping with the regulations at present governing co-operative societies in Israel. The mother-society may exercise its veto if any decision taken by the primary society appears to the representative of the former to be contrary to the principles of labour co-operation.

This, then, is the structure of ownership of the economic enterprises in *agrindus*-town.

(23) Should it prove necessary, because of existing circumstances, outside pressure, or development that has been undertaken prior to the creation of *agrindus*, compromise solutions, both of a short-term and long-term character, may be sought in a number of directions. Even modifications of the structure outlined in Paragraph 22 may be considered.

The situation in Israel in the years 1949–50 may be recalled to clarify this point. In that period the mass influx of immigrants, the desertion by the Arabs of many of their villages during and as a result of the War of Liberation, brought about a deviation from accepted principles of agrarian and social planning in this country.

(24) Both the *kibbutz* and the *moshav* had been based upon and had always observed the principle of self-labour. In the *kibbutz*, indeed, this principle is observed totally not only in agriculture but also in the various services. Thus even those not engaged in cultivating the soil, such as the shoemaker, tailor, carpenter, book-keeper and telephone operator, are also full-fledged members of the *kibbutz*. Even the doctor and teacher, whose salaries are paid by the Workers' Sick Fund and the Education Centre, respectively, can be members of the *kibbutz*, and even if they are not they can live there. But neither of them are employed by or receive their salaries from the *kibbutz*. The *kibbutz* pays its taxes and the authorities are obliged to extend to it and its members health, educational and other services. Members of the *kibbutz* who are employed as teachers or doctors put their earnings into the collective pool.

In the *moshav* the situation is different. The members are farmers who work their family holdings. Hired labour is not permitted. But the *moshav* as a corporate body maintains its diverse services through employees who are known as 'public workers'. These are paid according to the wage or salary rates of their respective trade unions. This category includes the employees of the creamery, the water works, the marketing society, the co-operative store, the secretariat and

20

the book-keeper. These, together with a number of independent artisans living in the *moshav*, form the class known as 'residents'. The artisans—shoemaker, tailor and others—work on the same basis as their counterparts in the towns.

In 1949 with the onset of an unprecedented wave of immigrants the national authorities, various public bodies and the labour movement—to which both the *kibbutzim* and the *moshavim* are affiliated—appealed to the latter to deviate 'temporarily' from their principle of self-labour and to give the new immigrants any employment available on the farms, in the factories, in building and in public services. The newcomers were housed in *maabarot* (transitional work camps) in the vicinity of the agricultural settlements. The response was general and immediate. Social mishaps arising out of the employment of hired labour were not wanting. Some *kibbutzim* have discontinued the practice, but the majority continue to employ hired labour, though they are seeking—up to the present, in vain—a satisfactory solution to the problem.

(25) One suggestion has been that some national or regional contracting agency assume responsibility for the employment of outside labour in the collective and co-operative settlements. The reform proposed, of course, is of a formal rather than a basic character. It has not been accepted.

The *agrindus* holds out a satisfactory prospect for the abolition of these categories of hired labour, through the transfer of many functions previously performed in the villages to the sphere of regional co-operation, the operative institutions of which will be concentrated mainly in the regional town. Should it at any time prove necessary to employ workers in a village in which they are not members, the task can be undertaken by one of the regional institutions without any direct relation between the village in question and the employees, in matters of wages and working conditions, being necessary.

All people who, not being members of any village in the district, are temporarily employed in the town's enterprises will be paid in keeping with their unions' rates. Permanent workers, however, should, in so far as this is possible, form co-operatives ranging round the enterprise in which they are employed. The societies they form will be allowed membership in the co-operative of that enterprise. In view of the fact that these workers forming co-operative societies on the lines of an *artel* will live in the regional town, we shall distinguish them by calling them 'residents'. The representatives of the 'residents' co-operative will take part in the management of the plant in the same way as the representatives of the employees' co-operative. The 'residents' co-operative will hold a special share in the plant

21

co-operative. Any member of a *kibbutz* or a *moshav*, who for any reason leaves his village and wishes to work and live in the regional town, will be at liberty to do so provided he joins the 'residents' co-operative of the enterprise in which he is employed.

The Co-operative Centre will assume responsibility for teaching the principles of co-operation to the residents of the regional towns.

The system proposed may be regarded as of a transitional character and a final solution left for a later phase of development. The younger generation, persuaded by their co-operative training—which must begin in school—and the co-operative environment in which they are brought up, may seek full membership in the *agrindus* by joining one of the constituent co-operative villages.

(26) We have already indicated that the salaries and wages of the residents will be in keeping with the scales of their respective unions. In Israel, certain co-operatives have instituted the principle of equal earnings. This principle, however, is practised in co-operatives in which the work involved is of a more homogeneous character, such as in the transport of passengers and freight. It may not be suitable for undertakings in which the tasks and responsibilities are more heterogeneous. Here some scale of earnings, based upon professional grading or skill, should be adopted.

Equalitarian principles may be allowed certain play through funds for insurance, pensions, vacation and education facilities, etc. Profit-sharing schemes can also be introduced.

In this context we cannot favour the basic principle of the *kibbutz*— from each according to his capacity, to each according to his needs. In the *moshav* this principle is not practised, and indeed the *kibbutz* is the only society in the world in which it is. *Agrindus*, we must stress, is not a commune but a co-operative.

Equal pay will be paid for equal work, but according to trade union rates. It is not irrelevant to point out that members of *kibbutzim* employed elsewhere are paid according to current rates in their profession or trade, but put their earnings into the common *kibbutz* pool.

(27) The residents of *agrindus*-towns (in the majority of which it may be assumed there will be residents who are not members of the villages in the district) must be housed in the co-operative estates of the type common in Israel and in other countries. The co-operative housing estates will be governed by their own statutes.

(28) In order to safeguard the co-operative character of *agrindus*-town economic initiative will be controlled. The town will promulgate legislation to meet its specific need. The essential purpose of *agrindus* is to preserve agriculture, the rural society and way of life. The method proposed involves the integration of industry and non-agricultural

occupations. Nevertheless certain bounds must be set to the non-agricultural development of the regional town. Our purpose would not be served if we permitted unguided and unrestricted development, jeopardizing the equilibrium that must be preserved between the rural district and *agrindus*-town. In the latter there will be no room for a private market or for cafés. Co-operative stores and restaurants will adequately fulfil this need. As far as possible, co-operative societies will engage in other trades and services as well.

It is by no means an easy task to pre-determine the optimum relative size of the town and its surrounding district. The executive body of *agrindus* will have to keep a close watch and steer developments with a firm hand.

(29) On the other hand, however, no restrictions will be placed on the expansion of the villages in *agrindus*. The most desirable development, of course, would be for all residents of the regional town to join the villages as members, to engage in farming or non-agricultural occupations in the town and to have their homes in the villages. The growth of the villages and of rural society generally in this fashion can only have the effect of strengthening *agrindus*.

The unguided and uncontrolled development of the regional town can only undermine the basic character of *agrindus*. For this reason constant supervision, through laws, regulation and day-to-day control, will prove necessary to maintain the optimal relation between the town and the rural area. The primary purpose of industry in town must be to process locally produced raw materials. Factories not related to farming will be developed only in so far as this is necessary to absorb the surplus manpower of the villages.

(30) Private economic enterprise (factories, etc.) is not wanted in the regional town, as it may divert regional development from the principles of *agrindus*. This does not preclude the participation of private capital in a co-operative undertaking, especially in cases where the former is of a specialized character and the entrepreneurs have know-how and technical ability to invest. Such partnerships between private persons and collective bodies, based upon co-operative principles, will not affect the co-operative character of their joint project. Here, too, the residents should be brought in through the co-operative society, holding shares in the capital of the project and being represented on the board of directors.

(31) The capital-structure of the regional economic projects will not differ in any way from that of the villages. Every society (or company) operating any enterprise will issue shares to the participating villages and to the residents' co-operatives. Share capital, of course, will constitute only a fraction of the total investment and the operative capital of the enterprise. The main sources of capital will

be loans from Government, public development corporations and banks. Industrial and agricultural development policy, as practised in Israel today, is to lend the sponsors of any project approximately 75 per cent of the investment capital required. In other countries promoting industrial development a similar policy providing for approximately the same degree of official assistance has also been adopted.

Once *agrindus* is recognized as a settlement system the allocation made to every prospective settler in Israel out of the national funds through the Jewish Agency should include an additional sum of agricultural investment capital to finance the participation of the village in the regional economic projects. These sums will provide the share capital as part of the own capital of the society operating the project in question. Other investment capital will come from the Industrial Development Budget; there will be recourse to the commercial banks for working capital. A special division of the Joint Agricultural and Settlement Planning Centre will be set up to co-ordinate the operations of the various *agrindus* organizations, for example, to plan the location of the economic and service enterprises (cotton gins, canneries, cattle- and poultry-feed stores, etc.). The bodies represented in this division must include the Ministries of Trade and Industry and of Interior and public institutions interested in this sphere of regional economic development. The implementation of the *agrindus* must proceed according to a national plan.

This is the general outline. It is not a Procrustes law to which local factors and conditions must be made to conform. Quite the contrary; it can be adapted to meet varying circumstances, particularly in the degree of co-operation instituted. The main feature of the entire project is *agrindus*-town—its social character, its dimensions, the lines upon which it is developed.

The thirty-one paragraphs describing principles and characteristics of the *agrindus* are flexible enough to be adapted to different political, social and economic conditions in capitalist as well as socialist environments. The two main and fundamental principles of the *agrindus* are: (*a*) integration of agriculture, services and industries; (*b*) co-operation between neighbouring villages within a region.

There are perhaps very few capitalistic countries—industrial or agrarian—where we do not find more or less diversified forms of a free co-operative movement.

The transition from the existing forms of rural organization to the *agrindus* forms, which appear revolutionary, is a long-term process. We do not aim to cover the entire face of the earth with *agrindi*. But we suggest a new form to be introduced step by step wherever possible. It seems to me that in Israel there is fertile soil for this innovation, that some of the prerequisites already exist; but even in Israel a beginning should be made with only a few *agrindi* in the more suitable regions, say two or three in the North, and a similar number in the South. If they prove successful, more can be established.

It must be clear that paragraph 30, formulating the attitude towards private capital in *agrindi*, is in keeping with conditions prevailing in some regions in Israel, where we find enterprises based on partnership between co-operative societies, public institutions and private capital. Even in the ideal *agrindus*, as I see it, participation of capitalists is not excluded, provided that the majority of voting shares of the enterprise are in the hands of the co-operative societies together with Government and other public bodies, if any. This proviso is, I feel, fully justified under Israel conditions, as it is the policy of the Government to grant loans of up to 75 per cent of the capital investment in the development of new industries. It would be unfair, accordingly, for such an investor to gain control of the enterprise with aid of government funds. But variations are possible according to conditions prevailing in different countries. There can be many types of *agrindus*, from the purely co-operative to the extreme capitalistic. Might it not be possible to establish an *agrindus* in a country like the U.S.A.? As we shall see below (Chapter VI) a rubber factory established in Utah had a far-reaching influence on its agricultural environment. The intention was to halt the desertion of the land by establishing an industrial enterprise in which some of the local rural population would be employed. In this we come closer to the evolution of a specific American type of *agrindus*. In a similar way a British or French capitalistic, co-operative or mixed type of *agrindus* might emerge.

Implementation of the plan will depend upon the character of the national regime, the willingness of Government to give preference to co-operative movements and institutions, upon the degree of sympathy with which the idea is regarded, and above all upon the resolution and ability of those who build up and maintain the *agrindus*.

It is catastrophic for Mankind that, for the time being, at least, nuclear physics has taken a destructive turn. But nuclear physics, we still venture to hope, will help build nuclear society.

Agrindus can serve as the nucleus of a better society.

III

METAMORPHOSIS OF AN IDEA

OR almost two thousand years, since the Judeo-Roman War in the first century of the Common Era, the Jews cherished the hope of returning to their ancient homeland. It was an aspiration that assumed different forms in successive generations. It was incorporated in the Prayer Book, it inspired messianic movements and it drove leaders of the people and large numbers of members of religious sects—Cabbalists, Hassidim, Perushim—throughout the Middle Ages and also in the Modern Era to make a pilgrimage to and to settle in the Holy Land.

Descendants of the ancient Judeans lived on in the country, in the villages, but they were pitifully few in number, and were hard put to preserve their Jewish identity.

It was only in the nineteenth century, prior to the modern Return to Zion, that residents of Palestine and thinkers in the Diaspora began to link this age-old hope of a return to the land with the cultivation of the soil. Sporadic attempts, which proved unsuccessful with discouraging regularity, were made to settle Jews on the land in 1834, 1838, 1855, 1859, 1860 and later.

The seed that was to strike root, grow and bring forth fruit was the Mikveh Israel Agricultural School founded in 1870. Eight years later a group of Jews from Jerusalem founded the first proper farming village. Soon Baron Edmond de Rothschild was to adopt the idea of Jewish settlement in Palestine, and though his motives were philanthropic, and his methods patriarchal, before many years had passed there were a considerable number of colonies—*moshavot* in Hebrew—in the northern and central districts of the country. In the decade beginning in 1882 a wave of immigration of unprecedented dimensions brought 25,000 Jews into the country and led to the establishment of twenty-three villages. In this period a movement of quasi-national character, the *Hovevei Zion* ('Lovers of Zion'), was already active. Towards the close of the century Political Zionism, founded and led for eight years by Theodor Herzl, determined the organizational structure of the Movement through the Zionist Congress which convened for the first time in Basle in 1897, and which developed as the supreme institution of the Movement, operating through its organs in the Diaspora and in Israel. Another decade of accelerated immigration, bringing another 40,000 Jews and leading to the establishment of another fourteen villages, came to an end with the outbreak of the First World War in 1914.

This wave of immigration, henceforth known as the Second Aliyah—the First had taken place in the eighties—comprised mainly young men and women, graduates of the clandestine socialist movement of Europe—mainly Czarist Russia—for whom the return to manual labour, to cultivation of the soil, was not enough. Their ideal was the moral regeneration of Man through truth, equality, co-operation and comradeship. Today these principles illuminate the lives of 85,000 people living in 228 *kibbutzim*. This collective settlement movement is divided into a number of groupings each of which is affiliated to a political party. These groupings are also distinguished by differences, for example in the administration and structure of their settlements, or in their internal regimes. In the past there were also differences of an ideological character, concerning the optimal size of the settlement, whether it should be 'closed' or 'open' and whether it should be based exclusively on agriculture or engage in handicrafts and industry as well. The passage of years has blurred these differences; certainly they are no longer matters of principle. There are still *kibbutzim* which have not developed industrial enterprises, but, in theory at least, they are not opposed to such development. Most members of the *kibbutzim* belong to socialist parties, but there are also settlements whose members are religious. The members of one grouping of *kibbutzim* are affiliated to a non-socialist liberal party, notwithstanding the fact that they themselves live a socialist way of life.

In this period the *moshav ovdim* came into being. This type of village incorporates five major principles: national ownership of the land; self-labour; mutual aid; co-operative purchase of supplies and sale of produce; and equal farm-holdings in each village.

In the twenties an interesting attempt was made to establish a Labour Corps. This will be discussed at length in a subsequent chapter. Co-operative villages have assumed other forms, too, the most interesting of which is the *moshav shitufi*—an intermediate stage between the *kibbutz* and *moshav*, combining collective methods of production with individualistic consumption.

From the 1870's to the outbreak of the First World War the operations of the Palestine Jewish Colonization Association, founded by Baron Rothschild, of the Hovevei Zion and the Zionist Movement, the developments set in motion by the Russian Revolution in 1905, and to a certain extent even by the Young Turks Revolution in 1908, combined to form a pattern of continuity. Indeed it was precisely because of this continuity—because after the foundation of the Mikveh Israel Agricultural School the work was persistently carried on—that these later efforts proved more successful than those made in the first half of the nineteenth century.

The First World War threatened to destroy everything that had been built up in the course of many years of arduous work. During the war the Turks ruling Palestine embarked on a ruthless policy of obstruction, deportation, persecution and liquidation that was to last three long years. The cessation of hostilities, however, in the final stage of which the Balfour Declaration was issued, created the conditions for a new influx of immigrants and a more favourable field for initiative.

The Yishuv—the Jewish community of Palestine—that faced the grim prospect of the Second World War in 1939 was a well organized body, with a ramified structure of institutions, movements and economic, social and cultural enterprises. Had it not been for the catastrophe which engulfed the Jews of Europe in the early forties, the development of Palestine might have taken a different course. Certainly had Nazi Germany not been defeated and mass slaughter been the fate of the Jews of Israel as well, the national renaissance for which three generations had striven would have come to an abrupt end. Happily for all mankind, however, the Allied Powers were victorious and their victory held out a more hopeful prospect for the sorely stricken Jewish people. The War of Independence, and the establishment of the State of Israel in 1948, opened up new vistas. An immense wave of Jewish immigrants, far greater than any which had reached these shores before, made up from the survivors of the European holocaust and refugees from national, religious and racial oppression in many other countries, now inundated the country. Political, economic and cultural conditions were completely transformed. The newcomers who were settled on the land preferred more individualistic forms of farming and social organization. Hundreds of *moshavim*—but only a relatively small number of *kibbutzim* —were established. The main stream of the immigrants, however, made for the towns. At the same time new urban centres which, for the time being only, certainly, are known as 'development' towns were established in various parts of the country. Agriculture progressed rapidly and output reached unprecedented levels. In the early 1950's there were scarcities of basic commodities, and rationing and controls had to be introduced. Soon, however, local

28

farm production had overtaken domestic consumption of staple commodities; intermittently there were even surpluses and market gluts.

Hand in hand with the rapid technological advance of agriculture went the new policy of industrialization. The Government, and in its wake, public bodies, embarked upon a resolute policy of attracting industrial investment capital and of ensuring a wide distribution of industrial development, that would not by-pass the 'development towns'. This policy proved comparatively, though not sufficiently, successful.

Under these circumstances, it seems, a significant rôle is reserved in the development of industry in closer relation with agriculture—or better still in the integration of industry and agriculture—for the agricultural settlements, particularly the co-operative villages, the *kibbutz* and the *moshav*, which have accumulated an invaluable store of economic and organizational experience, and have achieved high cultural levels. Twelve thousand workers are already employed, more or less in equal proportions, in *kibbutz* factories and workshops. Towards the end of the past decade their annual output totalled IL. 81 million[1] in industry and IL. 30 million in handicrafts.

The *moshavim*—and some of the *kibbutzim*—have not gone in for industrial development. On the other hand, *kibbutzim* whose factories have grown to considerable proportions have come up against many organizational, financial and, particularly, social problems of some magnitude. Manpower shortages within the *kibbutzim* have forced them to have recourse to hired labour, which has had unfavourable internal repercussions, chiefly due to the violation of the principle of self-labour. Certain *kibbutzim* have already reached the stage where they would prefer the removal of the factories to some neutral ground in the near vicinity or to a district centre where they could be run as co-operative regional projects. Organization on these lines offers many social and economic advantages. At the same time, however, many *kibbutzim* are still interested in the establishment of factories in their own areas. This is especially the case with *kibbutzim* with no personal experience of the difficulties involved and the problems that may develop as a result.

The successful operation of industrial plants of a regional character, established by the concerted efforts of several neighbouring settlements, furnishes another indication of the prospects of the *agrindus* idea. These enterprises include, among others, lucerne meal mills,

[1] IL. = Israel pound. In the years 1948–52 (February 13) one IL. was equal to $2·8; from 1952 till 1962 (February 9) one IL. = 55·5 U.S. cents; since 9 February 1962 one IL. = 33·3 U.S. cents.

cotton gins, slaughterhouses, cold-storage plants, packing houses, creameries, curing and grading houses and garages. The transfer of enterprises from any *kibbutz* to regional partnership can be carried out if that *kibbutz* wishes it. What is important is that such a possibility exists.

Government plans for industrialization include a long list of enterprises which can be located in rural areas. Further study, however, will be necessary to examine the relative advantage of given vicinities, availability of raw materials, transportation facilities, technical know-how and the like. Nor should the difficulty that will be encountered in the choice of products and in the successive phases of implementation be ignored.

Our brief survey, we hope, has served to show that while undoubtedly much remains to be done we can embark upon the development of co-operative agro-industrial regions, which by vertical and horizontal integration can combine to form ramified economic, municipal and social complexes. These large units incorporating the characteristics of town and country, of industry and agriculture, will constitute the *agrindus* of the future.

Government development can also continue. Factories are being and will be erected in the larger cities. We have recently been informed of official plans to establish another dozen towns in the Negev. At the same time voices have been heard calling for the rehabilitation, reconstruction and industrial development of Israel's ancient cities, such as Tiberias and Safad. Efforts are also being made to stimulate the expansion of outlying settlements. Their leaders come to the towns in quest of ideas, projects and potential investors in industry. *Kibbutzim* have already established sanatoria and rest houses, and some are planning enterprises such as tourist clubs. These efforts, of course, are not confined to the local leaders. They are actively supported and indeed often initiated by national leaders, Government officials and a variety of public bodies, who are all constantly on the look-out for foreign investment capital and technological know-how. Investors are given generous facilities and wide latitude in planning and initiative, and this has very tangible results including a growing national product, more employment and an expanding skilled labour force. But, in the long view, these new undertakings do not dovetail nicely into overall national plans. The Israel Government is liberal in its support of any manifestation of initiative, but not enough effort is being made to discover or stimulate potential initiative, for the organized agricultural community which has such remarkable agro-technical and social achievements to its credit has not been drawn into the orbit of industrial planning. The very fact that not one of the three sugar refineries established in recent years is owned and operated by the beet growers through their regional associations indicates how all factors in this country—Government, planners and even the farmers' organizations—are far from the *agrindus* idea.

IV

G'DUD HA'AVODA
THE LABOUR CORPS

THE wave of immigration into Palestine after the First World War included many young intellectuals who had been influenced by the revolutionary socialist movements during and after the War. They were strongly nationalist, both emotionally and intellectually, and aimed to build up a Jewish homeland on a socialist basis. They were looking for a satisfactory synthesis.

Palestine, in that period, had not developed a capitalist system, like that found in the countries from which these new immigrants had come and in which the socialist movement had developed. The entrepreneur-exploiter, whom one was in duty bound to fight, had not appeared on the social and economic scene. Socialist aspirations, accordingly, could be expressed only through the creation of new economic and social forms.

In the years 1919–20 the main source of employment in Palestine was the construction of roads. The nature of this and other branches of public works was—and is—amenable to co-operative contracting, and called for the establishment of temporary camps in which the workers lived on a communal basis. These factors, reinforced by the youth of the workers, stimulated aspirations toward independence, and a sense of a pioneering, social and national mission. In the summer of 1920 the *G'dud Ha'avoda*—the Labour Corps—was founded in a camp on the Tiberias–Safad road. This was the first road being built by pioneer-workers. Their camps served as the cradle of the *G'dud*.

The nine years of the *G'dud* fall into three distinct periods.

(1) The foundation years, when it engaged in public works, mainly road-building.

(2) The establishment of permanent agricultural settlements and the organization of labour groups in the towns.

32

(3) The decline of the *G'dud*, which ended, towards the close of 1929, with the merger with Hakibbutz Hameuchad.

It must, however, be noted that Hakibbutz Hameuchad grew out of the *G'dud*, when Ein Harod broke with Tel Joseph in 1923, thereby cutting itself off from the *G'dud*. It was in Ein Harod that the idea of the large, open *kibbutz* developed and Hakibbutz Hameuchad was born.

The *G'dud* distintegrated after nine years of existence but its work had not been in vain. It was the legitimate parent of the *kibbutz* movement; it constituted the prototype of collective social organization.

The principles of the *G'dud Ha'avoda* were as follows:

(*a*) Organization of disciplined workers' groups to be at the disposal of the General Federation of Jewish Workers (Histadrut) in all matters pertaining to national defence and employment.

(*b*) Establishment of a Workers' Sick Fund to serve all members.

(*c*) As far as possible to achieve self-sufficiency in meeting all needs of members of the *G'dud*.

(*d*) Surplus income to be utilized for economic expansion and improvement of working conditions.

(*e*) Reinforcement of the Federation of Labour and shaping of its expansion.

(*f*) Contracting for public and land improvement works, under its own management and upon its own responsibility. Members would also be despatched to undertake other tasks as required by the Federation of Labour.

(*g*) Development of various economic branches to meet its own needs and the establishment of settlements to be maintained permanently by its members.

(*h*) The *G'dud* must be organized on the basis of units, in keeping with places of employment.

(*i*) The units engaged on larger-scale projects must be split up into groups, for the purpose of undertaking given tasks of the project, or an entire branch upon the responsibility of the group.

(*j*) All affairs of the *G'dud* to be administered by a Council elected at a general meeting of the local units. The Council is the sole authorized representative in all external and internal affairs in respect of the location of units, budgetary matters and payment of taxes. Other institutions of the *G'dud* are: the Central Executive of the *G'dud* and the Local Executive of the unit. The Council must meet once every three months.

(*k*) Every unit must give an accounting of its production to the Council of the *G'dud*. After the issue of three warnings on the part of the Executive the Council may disband an uneconomic unit and distribute its members among other units.

33

(*l*) All income and payments received for the products of all units to be placed in the general fund of the *G'dud*.

(*m*) The Supreme Council of the *G'dud* allocates the income of the *G'dud* to meet the needs of members, economic expansion and the like.

(*n*) The Executive must submit to a meeting of the Council the balance sheet of each of the units and of the *G'dud* as a whole.

(*o*) Every worker in the country subscribing to the aims and the principles of the *G'dud* may be accepted as a member of the local unit of the *G'dud*, after a probationary period of one month and the formal approval of the Supreme Council.

In the course of time the *G'dud* declared its main objective to be agricultural settlement and the development of large, diversified farms. It did not, however, relinquish the idea of urban settlement. It regarded units based permanently upon wage labour in the private economy as feasible. Through the local units it sought from the very outset to penetrate into private industry.

The overall commune embracing both urban and rural units would—so it was hoped—ensure both mutual aid between the units and a close relation between production and consumption within the framework of the *G'dud*.

One proposal submitted envisaged one-third of the units supporting themselves by agriculture and two-thirds by outside employment. The latter were to assist in the establishment of agricultural settlements.

The proximity of factories and farms, according to the fathers of the *G'dud*, was to stabilize income and labour requirements in the agricultural economy.

FINANCIAL AND BUDGETARY METHODS

Social principles shaped the financial methods of the *G'dud*. An equal standard of living for all units was a sacred tenet.

In the initial period methods were haphazard and not uniform. There was considerable disparity between *per capita* living expenses in the various units, aggravated by the fact that the members had virtually no experience in budgeting, book-keeping and management.

In the course of time certain rules were laid down, an expenditure budget was authorized, and the treasurer required to manage the funds accordingly. Income was properly estimated. Expenditure provided not only for food, rent and transport to and from work, but also for education and care of children, the nursing of chronic invalids, aid to relatives, and a reserve fund.

The central treasury audited all income and expenditure. Even when earnings exceeded estimates, outlay was kept in check, both in order to build up a reserve to meet deficits incurred by any of the units and also for investment in new projects.

At the beginning of each month every unit was required to submit a report and a detailed abstract of its financial operations to the central accountant. The degree in which each unit had gone beyond or had lagged behind the budget estimates was duly noted. In this way close track was kept of the activities of every one of the units.

A far more complex task was the maintenance of an equal standard of living in the various units and the distribution of income.

The problems posed by maintaining equal standards and of carrying out works under fluctuating economic conditions—an unstable labour market, and changing wage levels and a fluctuating relation between gainful employment and work in the domestic and other services—gave rise to considerable controversy, which was aggravated following the departure of skilled workmen, who had been trained at the expense of the G'dud.

The financial administration also went through a number of phases. In the first eighteen months of the G'dud the income of all units was paid into the central treasury. This method broke down owing to the lack of leaders with the necessary financial experience and ability, and also because members were not prepared to make an extra effort to increase output.

Indirect methods of rewarding more efficient units by improving their living standards and of withholding similar rewards from inefficient units developed. Excessive concentration of authority once again led to disputes; decentralization and more financial autonomy for local units were demanded. The decision taken by the kibbutz of Ein Harod to break away from the G'dud led to the introduction of new methods.

In the second phase a greater measure of autonomy in financial and economic affairs was permitted. In fact the central treasury confined itself to supervision of income and expenditure of the Central Committee and the collection of dues. During this period a decision was taken to disband any unit that was not self-supporting.

The net result of these measures was a greater degree of local initiative. At the centre the accent was now on instruction and guidance, centralized purchasing, meetings of treasurers, accountants, organizers, skilled workmen and the like. In the units living standards rose but the equality which they had striven to maintain was now weakened.

In the third phase the G'dud—which was now registered as a limited company—introduced a system of placing bills on the market. The Central Committee issued a given amount in bills to

35

every unit; these bills served as working capital and the income of the units was used to redeem them! Thus while the units could decide freely on their expenditure, use of the bills was guided and regulated by the Central Committee. In this way adequate control over the units could be exercised from the centre. This method proved fairly successful but the G'dud was approaching its demise.

In all three of the phases referred to repeated efforts were made to make the units self-supporting by stimulating local initiative and by training members, and also to ensure the same standard of living. Leaders of the G'dud soon found that it must set up a large central reserve to assist in keeping living standards in the various units, working under disparate conditions, on a similar level.

The report of one of the controllers, dated 9 January 1924, includes a detailed and objective account reflecting the financial difficulties of the G'dud in that period. At this time, the report noted, the G'dud was saddled with the accumulated deficits of previous years besides lacking the working capital to finance its operations. However, the affairs of only four settlements, with 465 members— and none of the mobile units—were reviewed. These included:

Kfar Gileadi where 60 members were employed in agriculture and 10 in the construction of farm buildings. Tel Joseph, where there were 230 members but the farm could only provide employment for 170. The remaining 60 members found casual employment in drainage works and the construction of wooden huts for residential purposes, sometimes going as far afield as Haifa. Haifa itself had a small group of 23 members who were hoping to be reinforced by new immigrants. They worked mainly in the quarries and making concrete blocks for the Nesher Cement Factory. In Jerusalem there were 140 members employed as quarrymen and masons on road and building construction, in metal and carpentry workshops and in transport. Some of the women were employed in nurseries and in parks.

The controller's break-down of the various sources of employment were:

Gainful employment	52·5 per cent
Supply	16·5 per cent
Administration	4·5 per cent
Unemployment, illness and leave	26·5 per cent

These were the four most important bases of the G'dud and they held out longest.

Kfar Gileadi and Tel Joseph are today important collective settlements affiliated to the Ihud Hakvutzot Vehakibbutzim. Their respective populations in 1960 were 701 and 621 persons and their outputs were valued at IL.2·9 million and IL.2·1 million.

36

Ramat Rahel in the Jerusalem highlands is a different case. Its farming operations are restricted and it supports itself mainly by outside work in the capital.

Some mention should be made of the units long since disbanded. Plugat Migdal was the first unit of the *G'dud* to be founded, on the Tiberias–Semah road. When it undertook the construction of the Tiberias–Tabgha road it had 300 members, split up into 25 groups. In the main camp were the various services—the kitchen, shoemaker, laundry, sewing shop—in which the women were employed. Later a vegetable garden and a fishing group were developed.

A breakdown of 72,414 man-days of employment made at about this time is interesting.

Gainful employment	50 per cent
Supply	15 per cent
Administration	10 per cent
Illness	18 per cent
Rainy weather and leave	7 per cent

The high proportion of the labour force diverted to the domestic and other services led to the introduction of various methods to raise labour efficiency, such as the fixing of quotas and publication of noteworthy achievements. A special effort made to cover the deficit incurred in the early period proved successful and a small profit even remained.

The Migdal unit had a special function as the initial nucleus of the *G'dud*. Groups of members were periodically despatched to undertake tasks in other parts of the country. One of these reinforced the settlement of Kfar Gileadi which had been destroyed in the riots of 1920. Another group was sent to Beersheba where its members were employed on the repair of motor cars and lorries. A third, sent to Rosh Ha'ayin, was to serve as the nucleus of a unit to work on railroad construction work. A fourth, after a period in Haifa to learn the building trade, was stationed in the *moshava* of Karkur.

Once road construction work in Lower Galilee was completed the Migdal unit broke up into groups, some of which settled on the land in the Valley of Jezreel; one was despatched to reinforce the Jerusalem unit and a number to join units elsewhere. Indeed a number of other units developed out of the Migdal unit. Fifteen members of the latter formed the Rosh Ha'ayin unit to engage in railroad construction. It soon grew into a force of 250 members with its basecamp near the ancient stronghold of Antipatris. After the riots of 1921 the unit moved to nearby Petah Tikvah where the various services were reorganized. A sewing establishment, which had been set up here, proved very successful and was soon executing

37

outside orders. The unit also went in for poultry-breeding. When the work on the railroad was done this unit, too, split up into groups, most of which joined the settlements in the Valley of Jezreel.

So it will be seen that in a sense the years of major public works served as an incubation period for groups which later settled on the land.

TEMPORARY UNITS

Unemployment which set in after the road building period compelled the *G'dud* to seek out any work there was to be had. Attempts were repeatedly made to penetrate into various skilled trades. Temporary units of this type remained in existence until 1923. (It is noteworthy that in the course of its comparatively brief existence the *G'dud* undertook work in no less than forty-four areas in Palestine.) At the same time the *G'dud* tended increasingly towards units of a more stable character and towards the establishment of agricultural settlements.

AGRICULTURAL UNITS

The acquisition of the lands in the eastern half of the Valley of Jezreel opened up new vistas for land settlement. Now the *kibbutz* settlers began to contemplate, in more realistic terms, the establishment of a large collective settlement in which *agriculture would be combined with industry and handicrafts.* This had been the goal of Ein Harod, which had broken away in 1923 from the neighbouring settlement of Tel Joseph after an association lasting two years. Tel Joseph remained a constituent unit of the *G'dud* while Ein Harod served as the initial nucleus of Hakibbutz Hameuchad.

In 1921 the village of Kfar Gileadi joined the *G'dud*, and in 1924 its neighbouring settlement Tel Hai followed suit, linking up with the former. A number of other settlements (Beit Alfa, Hefzibah) entered into talks with the *G'dud*, with a view to joining it. These negotiations, however, proved abortive.

THE MARBLE QUARRY

A scheme to exploit deposits of marble in the vicinity of Metullah, only a few kilometres from Kfar Gileadi–Tel Hai, furnishes an interesting example of an attempt to develop a non-agricultural enterprise for these two settlements situated far from the markets for their farm produce. The *G'dud* acquired the concession for exploitation of the deposits and opened its quarry in 1925. The quality of the marble—of which many colours were available—was good and the local market extensive. Indeed the sole source of competition was the imported white Italian marble. The prospects for the export of coloured marble also seemed favourable. However, from the very outset, unexpected difficulties cropped up. The machinery acquired was obsolete, the *G'dud* had insufficient capital of its own to finance the enterprise fully and it sought private partners. Members of the *G'dud* worked on a wage basis. The main obstacle encountered was the lack of know-how. Serious errors led to deficits until the company was liquidated by Government order.

THE JERUSALEM UNIT

The group in Jerusalem constituted the largest and most highly developed urban unit of the *G'dud*. The idea of its founders was that it should penetrate the quarrying, stonemasonry and building trades. The initial phase of this unit was marked by haphazard—or rather complete absence of—planning. The members were devoid of any experience in organization or financial administration. Once, however, it had succeeded in acquiring a site of its own it began to operate in four different directions: casual employment; permanent employment; the development of independent workshops and factories; and the establishment of supply institutions. The unit was particularly interested in contracting, and indeed the *G'dud* undertook various building jobs, developed quarries and masonry works. Workshops established included a metal and carpentry workshop, and a smithy. Here again the main difficulties encountered arose out of lack of technical expertise and capital. The women were employed on tile-laying and stone polishing which, however, showed a loss.

The most pressing problem of all was vocational training. Many of the members left the *G'dud* after they had mastered a skilled trade at its expense. Even signed undertakings proved of no avail. Mechanization, some thought, might serve to check desertion of the *G'dud*, as machine-minding required specialized training besides offering a solution for less robust members.

After political differences had split the *G'dud*—and also the Jerusalem unit—in two, the latter continued to exist, thanks only to a group that settled at Ramat Rahel, near Jerusalem. The new settlement, however, joined Hakibbutz Hameuchad.

The Haifa unit which was founded in 1923 worked on road construction, and on the production of concrete blocks for the Nesher Cement Works, besides operating two quarries on its own account. Later the unit was employed in building construction, filling the salt pans at Athlit. The unit incurred successive deficits, and was subsequently disbanded as a result of political crisis.

The Tel Aviv unit, on the other hand, proved economically successful as the rapid pace of development in the young city kept its members fully employed. It undertook large building contracts, and its members enjoyed a higher standard of living than was the case in other units of the *G'dud*. This unit, however, also suffered from political differences and was dissolved.

IDEOLOGICAL DIFFERENCES

At a very early stage differences about the way of life in the *G'dud* soon developed into an acrimonious controversy. Two main trends emerged, one favouring the co-operative and the other the communal form of organization. The former insisted upon wages being paid to each worker individually, the allocation of supplies on a co-operative basis, a fair distribution of employment and a limited budget for personal expenses, the amount to be decided by the group. The latter called for a collectivistic way of life and for all income to be paid into a single national treasury.

The debate first took on a sharper note when the question of the connection to be maintained between the agricultural and the urban units was discussed. Some members of the *kibbutz* of Ein Harod demanded full autonomy in running their farm, in spite of a decision already taken by the Central Committee to the effect that in regard to funds for the purchase of supplies, agricultural settlements and all other units should be subject to the same rules. Over this question the settlement split, Ein Harod seceding with 105 members and Tel Joseph remaining with 225.

The General Federation of Labour (the Histadrut) sided with the dissidents. There had been friction already between the Histadrut and the *G'dud* over the standardization of wage rates and also because the Public Works Office of the Histadrut had refused to issue to the *G'dud* contracts without its supervision. These differences became more acute when a Communist group was discovered in the *G'dud*. It was only after the rift between the Right and the Left became absolute that relations with the Histadrut improved.

At the meeting of the Council of the *G'dud Ha'avoda*, held in Tel Joseph in 1926, the left-wing was formally expelled. A subsequent poll revealed that 294 members had remained with the *G'dud* while 195 joined the left-wing *kibbutz* which also bore the name of *G'dud Ha'avoda*. Sixty-five members remained neutral; the majority, however, returned later to the *G'dud Ha'avoda*. A smaller number afterwards joined Hakibbutz Hameuchad. Thirty-eight members in all decided to leave.

The left-wing *kibbutz* remained in existence only for a year. A few score of its members left for Russia as an organized group while the rest dispersed.

The *G'dud* continued its corporate independent existence for about three years after the break. In December 1929, a Council meeting to discuss the merger of Hakibbutz Hameuchad (Hebrew: The United Collective)[1] and the *G'dud* was held in the settlement of Yagur. The settlements of Tel Joseph, Kfar Gileadi–Tel Hai and Ramat Rahel now joined Hakkibutz Hameuchad. By this time the urban units had been disbanded.

In the course of its nine years (1920–29) the *G'dud Ha'avoda* had operated in forty-four different areas in Palestine. It had reached its peak in 1925 when it had 665 members, but it was estimated that 2,500 men and women had at one time or another served in its ranks.

We have dealt at length with this episode, which must be regarded as an unsuccessful experiment, because it constituted a phase in a process, the beginnings of which we have already considered in a previous chapter. This process is still in progress.

Certainly we might not have reached the present stage and many things might have developed differently had it not been for the *G'dud Ha'avoda* and the immense efforts made unremittingly over a period of nine years in the course of which the principles of co-operation, equality, of group living, of the combination of agriculture with industry, handicrafts, public works and national defence were reassessed and redefined.

Despite the initiative, the daring, the breadth of vision and the practical ability which characterized it, the *G'dud* failed to overcome the crisis in its ranks. It was eroded from within by an ideological ferment of a political character. Twenty-five years later Hakkibutz Hameuchad was also upset by internal political dissension, as a result of which a number of settlements were split in two. At this time, however, the *Kibbutz* was strong and emerged from the crisis strengthened, while the country was enriched by another four collective settlements. Hakibbutz Hameuchad, it transpired, was not immune to political crisis, but it possessed sufficient social, economic and organizational strength to withstand the shock.

The full title of the *G'dud* had originally been *G'dud Ha'avoda Vehahagana* (Labour and Defence Corps), but the last word has been discarded for tactical reasons. The function of defence, however, had always been its primary concern.

We have already indicated in the Introduction the importance we attach to continuity, to unbroken development. It was entirely due to

[1] For more detailed explanation of this and other organizations of collective and co-operative settlements see: H. Halperin, *Changing Patterns in Israel Agriculture*, pages 205–8.

a continuous process of development, the establishment of many new *kibbutzim*, and to the organization of defence in these and other settlements, that higher social, economic and organizational levels could be attained, the Israel Defence Army and the State of Israel could be set up, and that we can look to new and exciting achievements in our constant march forward.

This chapter has also served to indicate that the integration of agriculture and industry was taken in hand in Palestine as far back as 1920, long before short-lived attempts, resembling in certain respects the Palestinian experience, were made in Soviet Russia. A threefold integration—that is, with the addition of defence—was made about forty years before the establishment of the autonomous communes recently proclaimed in People's China. It is still too early, however, to make any assessment of this gigantic project which has not been tried in the furnace of experience.

The dramatic and checkered life of the *G'dud Ha'avoda* was not in vain. It was a necessary phase in a continuing process that is rising to higher levels under changing conditions. Whatever course future development may take the thread leads back ultimately to the trials of the *G'dud Ha'avoda* in the 1920's.[1]

[1] In the preparation of this chapter use has been made of the archives and collections of Beit Trumpeldor in Tel Joseph, of articles published in Palestinian weeklies of the period as well as personal knowledge and experience.

V

WORKERS' ENTERPRISE

To obtain a clear picture of the national economy one must have knowledge of such factors as resources and production, the balance of payments and foreign trade, private consumption, the volume of investment, prices and wages and manpower. Here, however, we are not dealing with Israel's economy, but with a more restricted and clearly defined subject. A cursory study of the table below is sufficient to prove that Israel is not an agrarian country—for only 17 per cent of its earners are employed in agriculture—nor is it an industrial country—for only 23 per cent are employed in industry, and it is very poorly endowed with natural resources.

Distribution of Earners—Economic Branches (per cent) 1960

Agriculture	17·1
Industry	23·2
Building and public works	9·2
Electricity, water and sanitary services	2·3
Commerce, finance and insurance	12·2
Transport, communications and storage	6·3
Public services	22·0
Personal services	7·7
Total	100·0

In the 1920's 76 per cent of the country's population lived in the rural districts while 24 per cent lived in the towns. Jewish immigration radically altered this distribution which today is reversed with 76 per cent in the towns and only 24 per cent in the villages.

The Arab sector accounting for 10 per cent of the population is still largely of an agrarian character. The Jewish sector's agriculture is based upon irrigation and is intensive in its labour and capital input, in its use of the most up-to-date agro-technical methods,

43

machines, chemical fertilizers and insecticides. It is served by a ramified network of farm schools, and institutions for agricultural instruction and training. Not many years ago Israel's agriculture supplied only a fraction of the country's needs. Today three-quarters of domestic consumption is met by local production. It is short of grains but has a surplus of livestock products (milk and eggs), vegetables and certain varieties of fruit. For many years, perhaps always, Israel will be a wheat-importing country. Any further expansion of agricultural production must be intended for export. So far, however, only one export commodity has been satisfactorily developed—citrus. No other which could be expected to be welcomed in overseas markets has so far been discovered, though experiments in this direction, on a comparatively restricted scale, are continuing.

Under these conditions a policy of active industrialization was inevitable, and the Government is granting very liberal concessions to foreign investors and is welcoming both public and private initiative in practically every field.

Over the past few decades workers' enterprises have developed, assuming very considerable proportions.

The General Federation of Jewish Labour in Eretz Israel (more commonly known as the *Histadrut*, viz. Federation), founded in 1920, is one of the most striking manifestations of economic and social life in this country. The creation of the Histadrut had been preceded by the Agricultural Workers' Organization set up in 1911, and important pioneering work in the establishment of various workers' institutions, organizations and unions conducted over a period of years. The first *kibbutz* had been founded in 1910. The Histadrut did not confine itself to labour organization, but embarked upon social and health insurance, establishing an extensive network of clinics, hospitals and convalescent homes, and also various mutual insurance and provident funds. It also became active in the educational sphere, catering for the needs of both adults and children, organizing Hebrew classes for adults and vocational training courses, and proved very successful in the field of co-operation. Side by side with certain of the highest forms of co-operation, without parallel in any other part of the world, more conventional co-operative enterprises and companies controlled and managed by the Histadrut through its representatives have been developed in transport, marketing, credit and banking, construction, public works, industry and handicrafts.

The population of the Histadrut—that is members together with their dependants—constitute 58 per cent of the inhabitants of this country. Members of the Histadrut make up 70 per cent of all earners in Israel. Wage-earners, indeed, constitute two-thirds of all employed persons in Israel, and 90 per cent of the wage-earners are organized in the Histadrut.

In 1958 the workers' settlements had 185,000 inhabitants, i.e. 12 per cent of the total population of Israel. Of these, 83,000 were in the *kibbutzim* and 102,000 in the various types of *moshavim*.

Workers' enterprises employ 172,000 persons, that is one-quarter of the Jewish labour force. More than half of them—90,000—are settled in the workers' villages which are the major source of productive employment within the workers' economy.

We shall discuss Histadrut enterprises briefly, after which we shall describe industrial development in the agricultural settlements.

After the Balfour Declaration and the conclusion of the First World War immigration into Palestine assumed very considerable proportions. The Mandatory administration became an important employer of labour, mainly through its large-scale public works projects—road construction and railway development—and also the erection of new buildings. The immigrants themselves created much employment with needs such as housing. Founders of the Histadrut were reluctant to leave building and public works exclusively in the hands of private contractors. By setting up its own contracting agencies, they felt the Histadrut would more effectively be able to safeguard the interests of the workers and ensure fair wages and social conditions. As far back as 1920 a Histadrut Public Works Office was set up; in the course of time this Office developed into the Sollel Boneh Contracting Corporation. The latter went through some difficult periods, especially during the recurrent economic depressions. Today Sollel Boneh and its subsidiary Koor are the main contracting bodies in Israel, their operations including many major undertakings in a series of foreign countries.

The Histadrut has another contracting body, Yachin-Hakal, which is concerned with agriculture—mainly citrus growing—and undertakes the planting and cultivation of groves for local residents and foreign investors, particularly the latter.

Both of these corporations have set up other enterprises, including factories.

Structurally Sollel Boneh is made up of three subsidiaries: the Building and Public Works Corporation; Koor, which operates in the field of industry, and the Overseas Works and Harbours Corporation. The Koor combine includes the Phoenicia Glass Works in Haifa, producing hollow and flat glass, the Phenikia Factory in Jerusalem, manufacturing precision instruments, an iron pipe works in Acre, a radio-crystals works in Holon (near Tel Aviv), quarries and concrete products works in Haifa, as well as a wide variety of metal workshops, joineries, garages, etc., to meet its internal requirements. The

45

Overseas Works and Harbours Corporation owns and operates the Ogen ship-building yard in Haifa Port.

Yachin has canneries and factories for fruit juices. Hamashbir Hamercazi—the national consumers' co-operative—has also set up factories and workshops, besides holding an interest in other important concerns. Among other enterprises it operates flour mills, bakeries, feed-mixing plants and textile works. Enterprises in which it is a partner include a sugar refinery, an artificial teeth factory, a fertilizers and chemicals plant, the Nesher Cement Works, and factories for the manufacture of products such as paper and welded steel pipes.

Tnuva—the national co-operative for marketing farm produce—has nine large regional creameries, fruit and vegetable canneries, a citrus products factory and an alcohol distillery. It also has an interest in a sugar refinery.

Workers' Enterprises, 1958

Number of workers employed, gross output, gross output
per worker (average)

Category	No. of workers		Gross output		Gross output per worker (average) IL.
	Number	Per cent	IL. 1000	Per cent	
Koor	7,077	25·6	114,034	29·7	16,613
Even Vasid	1,600	5·8	13,000	3·4	8,125
Other economic enterprises	3,858	13·9	99,589	25·9	25,812
Productive co-operatives	3,420	12·4	44,610	11·6	13,045
Workers' settlements	4,914	17·7	68,157	17·8	13,870
Total, Industrial enterprises	20,869	75·4	339,390	88·4	16,461 (average)
Internal enterprises	1,500	5·4	18,000	4·7	12,000 ,,
Workshops in settlements	5,300	19·2	26,500	6·9	5,000 ,,
Total, Industry and handicrafts	27,669	100·0*	383,890	100·0†	13,985‡ (General average)

* Rise in 1958 (1957 = 100)—107·5
† Rise in 1958 (1957 = 100)—112·5
‡ Rise in 1958 (1957 = 100)—104·8

Share of workers' enterprises in Israel's industrial complex in 1958:

In number of workers　20 per cent
In value of output　　22 ,, ,,

46

Other Histadrut enterprises concern the manufacture of products such as plastic goods, knitwear, electrical goods, sewing machines, bricks and ice, and work such as the operation of printing presses.

Then there are 146 co-operative factories and other productive enterprises including, for example, bakeries, printing works, joineries and cabinet-works, metal works, garment and footwear industries, building materials plants and garages.

Finally the workers' settlements—mainly the *kibbutzim* and the Regional Councils—own and operate 156 industrial enterprises.

The tables shown opposite and below give a breakdown of the number of workers, gross output and gross output per worker in major branches of Histadrut industry in 1958.

Workers' Enterprises, 1958

Number of workers, gross overall output, gross output
per worker (average) in main branches

Branch	Number of workers		Gross overall output		Gross output per worker (average) IL.
	Number	Per cent	IL. 1,000	Per cent	
Metal	4,610	16·7	61,188	15·9	13,273
Electrical goods	468	1·7	4,977	1·3	10,635
Cement and cement products	2,293	8·3	49,528	12·9	21,600
Glass and ceramics	1,499	5·4	16,401	4·3	11,187
Quarries and sandpits	1,768	6·4	14,885	3·9	8,419
Rubber and plastics	712	2·6	17,027	4·4	23,914
Shipbuilding	398	1·4	3,720	1·0	9,347
Food and tobacco	4,507	16·3	118,180	30·8	26,221
Chemicals	501	1·8	8,949	2·3	17,862
Timber	1,927	7·0	28,133	7·3	14,599
Printing and paper	677	2·4	4,964	1·3	7,332
Tanning and leather	278	1·0	2,950	0·8	10,012
Textiles	346	1·2	4,693	1·2	13,564
Garments	43	0·2	455	0·1	10,581
Miscellaneous	842	3·0	3,340	0·9	5,045
Total, Industrial enterprises	20,869	75·4	339,390	88·4	16,431
Internal enterprises	1,500	5·4	18,000	4·7	12,000
Workshops in settlements	5,300	19·2	26,500	6·9	5,000
Total, Industry and handicrafts	27,669	100·0*	383,890	100·0†	1 3,983‡

* Rise in 1958 (1957 = 100)—107·5
† Rise in 1958 (1957 = 100)—112·5
‡ Rise in 1958 (1957 = 100)—104·8

In the four years preceding 1958 progress was rapid. In this period the number of workers employed grew by 28·4 per cent and gross output by 66·8 per cent.

Structurally Histadrut enterprise differs radically from that of the private sector. Basic industries, such as non-metallic minerals, mines and quarries, metallurgical and machinery plants make up 44 per cent of the production of the former as against 23 per cent of Israel's industry as a whole. This is also true of the timber and plastics industries. The most important branch of workers' enterprise is the food industry, particularly that which is dependent upon agriculture and agricultural raw materials. Second is the metallurgical industry followed by cement and timber (and their products).

The distribution of workers in Histadrut factories and in Israel's industry is interesting:

Number of workers per enterprise	Israel's industry (per cent)	Histadrut industry (per cent)
–10	21·2	3·9
11–20	12·4	8·7
21–30	8·7	6·3
31–50	12·0	6·6
51–	45·7	74·5

So it can be seen that in Histadrut industry there is a distinct trend towards larger factories.

A further breakdown in the last category in the foregoing table—namely plants employing more than fifty workers, which account for 74·5 per cent of Histadrut industry—shows that enterprises employing more than 500 account for 26·2 per cent of the total. Their share of gross output is even higher—31·9 per cent. Average output per worker was also higher—over IL. 19,000, about 54 per cent more than the average in smaller plants with 11–20 workers.

The structure of Histadrut industry in terms of gross output and labour force will probably be of interest (in percentages). (See Table opposite.)

In other words the larger plants with a gross output of more than IL. 10 million produce a quarter of the total output. The factories with an output of IL. 5–10 million account for more than another quarter. In the years that have elapsed since the establishment of the State of Israel, Histadrut industry has shown far greater enterprise and activity than in the preceding twenty-five years. Existing plants have been expanded while many new ones have come into operation. Notable progress has been made in both production and marketing.

The labour force has increased rapidly, though recently Israel is beginning to suffer from a shortage of labour, particularly of skilled workers. Collaboration between the institutions operating in the industrial field, and between these institutions and the Government—especially the Ministry of Trade and Industry—has grown closer. One facet of this collaboration has been the establishment, jointly by the Histadrut and the Government, of the Tiuss Corporation, for the promotion of industry in the development towns.

Output	Plants	Workers	Gross output
Less than IL. 1 million	85·5	34·3	24·1
IL. 2 ,,	7·3	14·1	12·7
IL. 5 ,,	2·7	14·0	10·2
IL. 10 ,,	3·3	27·9	27·8
Over IL. 10 ,,	1·2	9·7	25·2
	100·0	100·0	100·0

Striking progress has been made in industrial development in the workers' agricultural settlements. There has, however, been a recession in urban producers' co-operatives.

In spite of the steady growth of privately owned industry, workers' enterprise—especially in basic industries and plants processing agricultural raw materials—has held its own and even registered a greater advance.[1]

[1] The balance sheets of many settlements as well as material published by the following institutions have been used in the preparation of this chapter: Economic and Social Research Institute, General Federation of Jewish Labour in Eretz-Israel, Executive Committee, Statistical Abstracts and Industry in the Labour Economy (facts and figures) 1959–60.

Other Sources:
(1) G. Zederovitsh. 'Labour Industries in 1957'. *The Economic Quarterly* (Hebrew) No. 20, 1958, p. 431.
(2) G. Zederovitsh. 'Trends in Development of Histadrut Industries'. *The Economic Quarterly*, No. 24, 1959, p. 412.

VI

INDUSTRY IN THE WORKERS'
SETTLEMENTS

Wᴇ have data on 126 industrial enterprises in the workers' settlements. These enterprises, which employ 5,000 men and women, have a total annual output of IL. 70 million. In addition there are a large number of workshops with about 6,000 workers which have an output exceeding IL. 25 million. The main branches of industry are timber (employing 30 per cent of all workers), metal and electrical goods (about one-fourth), and food (about one-fifth). These are followed, in order of importance, by cement, glass, ceramics, quarries, chemicals, printing, leather, rubber and plastics and textiles.

Judged by the number of workers these establishments are not large. About one-quarter employ up to 10 workers, another third have 11–20. A small number have 100–300 workmen, and no more than three have more than 300.

In terms of gross output: one-fifth have an output of under IL. 100,000; half of them are in the IL. 100,000–500,000 bracket; 17 per cent—under IL. 1,000,000. Five of these factories have an output of under IL. 1·5 million; 2 up to IL. 2 million; 6 up to IL. 5 million and 2 over IL. 5 million.

The *kibbutzim*, it seems, experience considerable difficulty in operating large- or even medium-sized factories, principally because of their manpower problems. The employment of hired skilled or unskilled workers not only runs counter to a social principle but gives rise to internal complications. There have been cases in which settlements have asked the regional council to take over and run on co-operative principles an enterprise within the boundaries of those settlements. It seems to us that in view of technological advances and the economics of scale favouring larger enterprises it will become increasingly advisable for the settlements to relinquish control of their factories.

50

The *moshavim* were opposed in principle from the very outset to the development of factories within their borders. The founding fathers of the *moshav* aspired to return to the soil, to cultivation of the land. Only twenty years ago they were still opposed to the mechanization of farming which deprived men of their work and estranged them from Nature. New developments and the reduction in many settlements and families of their 'surplus' earners have revived the idea of industrialization. The author has succeeded in convincing some of the leaders of the *moshav* movement, especially in the younger echelons, to support the establishment of regional industrial enterprises. Some of them are attracted to the idea of regional industrial development because of their fear that factories in the *moshav* will disturb the even tenor of agriculture or affect the agricultural principles of their movement. There are others among the younger generation in the *moshavim* who say, 'One ideological concession has already been made . . . when the leaders of the movement consented to the establishment of enterprises to process farm produce and were not opposed to members of the *moshavim* being employed in them. The condition, however, is that these projects be of a regional character, not connected directly with the *moshav*.' A spokesman of the younger *moshavim* has added, 'We must take another step. Instead of a *moshav* settler seeking a day's work elsewhere, why should he not work in his own enterprise, belonging to his *moshav*? The *kibbutzim* and the *moshavim shitufiim* are already doing so?' He goes on to propose three categories of non-agricultural enterprise which he considers suitable for the *moshav*: home industries after the Swiss and the Japanese model, projects owned and operated by the village and regional undertakings.

The *moshavim* have not adopted any one of these three alternatives. It would be advisable, accordingly, for them to opt for the third, to which the experience gained by the *kibbutzim* over a long period is leading. Let the *moshavim*, too, join existing or prospective co-operative non-agricultural projects in their neighbourhoods. The aspect they find most attractive is that the second or third son of the family who has no employment on the farm will be able to find work in industry and nevertheless continue to live with the family.

A study of the influence exercised by a factory established in an agricultural environment in the state of Utah (U.S.A.) should prove illuminating, and though of course, because of the wide disparity between conditions in this and that country, no analogies can really be drawn, there is much to be learnt from the American experience.

. . . 'In recent years, many industrial concerns have established plants in rural communities,' the authors of the study say. 'This

51

development affects rural people in several ways. It provides additional employment opportunities and income both for farm and non-
farm families in the rural area through the jobs provided in the plant
as well as in expanded service and trade establishments. This, in turn,
leads to an increase in the local market for food and other products.
In addition, human and capital resources that might otherwise leave
the area are retained locally.

When many people switch from farming to industrial work or to a
combination of part-time farming and a non-farm job, changes in the
economic and community life of the area are likely to result.

This report is based on data from interviews, taken in the spring
of 1958, with 205 employees of the Thermoil Rubber Company.
However, plant workers have moved more often than either farmers
or other non-farmers in the last ten years; farmers had made the
fewest changes in residence. Workers in the factory had more
experience in non-farm work than other non-farm workers or farmers
and had worked at more non-farm jobs than either of the other
groups.

. . . One fourth of the plant workers lived on farms and the
majority of them were farm operators. A like proportion of farmers
interviewed held non-farm jobs. These part-time farmers used more
hired labour, and generally raised crops of livestock requiring less
labour than did full-time farmers in the study. Farms operated by
plant workers averaged about one-half the average of those operated
by full-time farmers. There was a tendency on the part of both plant
workers and other non-farm workers living on farms to reduce average and increase number of livestock after taking an off-farm job.
However, an increase in livestock was also noted among full-time
farmers. After taking non-farm jobs, farmers had reduced the number
of days worked on their farms per year by one-half or more.

The majority of plant workers and other non-farm workers who
were also working on farms preferred non-farm to farm work. However, only about one in twenty indicated he planned to quit farming.
The majority of non-farm workers stressed economic security as a
reason for continuing to work their farms.

Steady work and adequate pay were rated as most important job
conditions by all groups of respondents.

The majority of all respondents expressed favourable opinions
towards the factory and its influence on the community. The factory
was considered to have played an important part in stabilizing the
population of Nephi, the town in which it is located.' [1]

[1] 'Industrialization and rural life in two central Utah counties' by John
R. Cristiansen, Sheridan Maitland, John W. Payne. Bulletin 416. Utah
Experiment Station, Utah State University, Logan, in co-operation with

The problem we have to contend with differs from that encountered in Utah. There the objective was to strengthen the town of Nephi by stabilizing the agricultural environment through the development of part-time farming in which both townsmen and villagers would engage. It is doubtful whether this is a long-term solution in view of the conditions of American agriculture. Mechanization and techno-logical development generally will require robust specialized farm-ing, and it remains to be seen whether indeed the farmers' attention can be properly devoted to agriculture and industry at one and the same time. It might be done provided the various functions and jobs are split up among the members of a family—one going in for farm-ing, another taking some non-agricultural job and so on.

Our purpose in citing the above, notwithstanding the widely dis-parate trends, was to underline the fact, at least by one other example, that this problem of integrating agriculture and industry is confront-ing many regions, peoples and states.

The distribution of industrial enterprise in the workers' agricul-tural settlements on the basis of grouping affiliation and region is interesting.

Hakibbutz Ha'artzi of Hashomer Hazair (Hebrew: The National Collective of the Young Guard), which in the beginning insisted on agriculture to the exclusion of all else, today operates forty-eight factories. Hakibbutz Hameuchad, which from the very outset favoured the integration of factories and workshops in its settle-ments—carrying on the tradition of the *G'dud Ha'avoda* from which it developed—has forty-six. The Ihud Hakvutzot Vehakibbutzim (Hebrew: Federation of the two forms of collectives: *kvutza* and *kibbutz*) has thirty-nine, while there are another twenty-three indus-trial enterprises in other settlements. The latter include factories operated by religious and non-affiliated *kibbutzim*, regional under-takings as well as a small number in *moshavim*.

Each of the three major *kibbutz* groupings has adopted a different policy towards industrial development.

Hakibbutz Hameuchad, the first advocate for the introduction of industrial enterprise when other movements favoured only agri-culture in their settlements, has not yet resolved the question whether it should establish some central movement institution to operate in the field of industrial development. The affiliated *kibbutzim* are afraid of the restrictions such a body might impose upon their individual plans for development. This movement, it should be noted, has

the Brigham Young University and the U.S. Department of Agriculture. 32 pages.

effective central institutions for other spheres, including finance, building and supply.

The forty-six factories of Hakibbutz Hameuchad can be classified in ten main categories. A number of them have already acquired a considerable reputation in the manufacture of irrigation equipment, agricultural machinery and kitchenware (e.g. Givat Brenner, Yagur, Beth Hashitta, Ein Harod, Naan, Kefar Szold, Ashdot Yaacov and Menara), timber (again Givat Brenner and Ein Harod, and also Alonim and Givat Haim), tinned fruit and vegetables (Sdeh Nahum, Givat Haim, Beth Hashitta and Givat Brenner), etc.

In the past some members of Ihud Hakvutzot Vehakibbutzim were strongly opposed to the introduction of factories, and certain settlements do not go in for non-agricultural enterprise to this day, though no longer on grounds of principle. The Ihud has an efficient industrial department. Some of the factories in its settlements are of very considerable proportions by all standards, such as Afikim's Kelet Plywood Factory, operating mainly for export, whose overseas sales last year topped two million dollars. Other important plants in Ihud settlements are: the Mehaprim Works in Kfar Gileadi, producing hydraulic excavators and loaders attached to tractors; the Taal and Gazir Works in Mishmarot and Gezer respectively, both of which manufacture bruce boxes. Taal also produces plywood, its exports totalling one million dollars last year. Kfar Hanassi manufactures irrigation fittings and accessories as well as other non-ferrous goods —aluminium, bronze and brass. Agricultural machinery is made by Tel Joseph and Yifat. A number of settlements have canneries.

Hakibbutz Ha'artzi has developed a special system of *kibbutz* industry. In addition to the local enterprises of *kibbutzim* whose products include machinery and agricultural equipment, chemicals, plastic, rubber and canned products, this movement has a registered company, Mifalei Techen, Ltd., combining all its factories. This company engages in the supply of raw materials, financing and in the marketing of the finished products. It also furnishes technical, organizational and economic guidance. A special feature of the movement is the operation by this Company of three plants—the Askar Paints and Chemical Works (attached to Kfar Masaryk and Evron), the Naaman Pottery and Fireproof Brick Works, and Galam Works (attached to Kibbutz Ma'anit) whose products include glucose, starch and cornflower—owned co-operatively by all members of the movement.

All of these enterprises have a large turnover.

What, from the point of view of the *kibbutzim*, are the advantages and disadvantages of operating industrial enterprises within their own borders?

The advantages are as follows:

(1) Agrotechnical progress and rising levels of productivity have caused and will continue to cause a reduction in the labour force required in agriculture. The busy seasons in which a larger number of hands are necessary are much shorter than the quiet seasons when the members of the *kibbutzim* are under-employed.

(2) Higher yields have already resulted in gluts in certain branches of agriculture. Income has dropped. Industry can bring in a higher income and thereby even out the general level of income in the *kibbutz*.

(3) The *kibbutzim* dispose of considerable technical, organizational and administrative talent. Many of their members have been seconded to important positions in their respective movements, the Histadrut and the State. It is in the interest of the settlements to provide these members with suitable outlets for their special abilities at home.

(4) Many members of the younger generation find little satisfaction in farming. They show a preference for technology and engineering. They are sent to technical schools, but if they do not find suitable employment at home when they graduate they may wander further afield. Factories within the settlement could provide a congenial sphere for them.

(5) Because of age or their health certain members have to work in an enclosed area or to do work requiring less physical effort than is necessary in agriculture. Many branches of industry could solve this problem.

(6) Management and overhead expenses are lower than in the towns. The same is the case with the living costs of a rural worker (enjoying the same standard of living). Buildings to accommodate plant and personnel are also cheaper.

(7) In the village the factory or the workshop belongs, integrally, to a larger body, united for consumption, agriculture and industry, living in a diligent and enterprising environment and possessed of ramified connections throughout the country.

(8) The village can help in the promotion of urban branches of industry both as a supplier of raw materials and a consumer of the finished products.

Problems and Disadvantages

(1) Hired Labour. The medium-sized *kibbutz*, normally speaking, cannot furnish much labour for an industrial enterprise. Every factory calls for permanent and specialized personnel. Members laid off from farm work during the quiet seasons can only undertake functions of secondary importance in the factory. Once the factory

A.—E   55

reaches the stage where it requires forty to fifty permanent workers it must have recourse to the outside labour market and grave social problems in the settlement result. The older members adhere firmly to this sanctified principle of self-labour and are prepared to forgo industrial enterprise if the latter should undermine it in any way. The younger people, however, possessed of a more economic and techno-logical turn of mind, are beginning to feel that there is no escape for the *kibbutz* from the process of industrialization and that it should be so constructed that the members work in agriculture and that hired labour be employed in the factories.

(2) Frequently when a *kibbutz* is contemplating expansion or even the launching of an industrial project based upon more complex processes, it cannot find the necessary technological or management personnel among its own members. It is virtually impossible to enlist such personnel outside.

(3) There is an acute shortage of investment and working capital. The settlements have, in any case, to contend with major financial problems. Industrial development requires large sums which cannot be raised, particularly in view of the existing obligations of the agricultural economy.

We believe that the regional system, that is the co-operative maintenance of industrial and other non-agricultural projects by a larger number of neighbouring villages, can provide a satisfactory solution. The advantages of combining industry with agriculture, as already outlined, are retained in regional enterprises and contribute to the benefit of the individual settlement as well. The difficulties, however, are overcome and the disadvantages removed.

The regional industry can employ hired labour without detriment to the *kibbutz* society. Co-operative forms may, of course, be devel-oped, extending membership in the regional enterprise to workers who do not belong to the *kibbutzim*. In other words the co-opera-tive society running the enterprise may comprise both *kibbutz* settlers and industrial workers living in the neighbourhood.

Association of twenty or thirty neighbouring settlements also offers a prospect of solution to the second of the disadvantages listed. Firstly the chances of enlisting local managerial talent are corre-spondingly larger, while a regional project can prove more attractive for outside specialists, engineers and managers.

Finally, a larger body, comprising a considerable number of agricultural societies, will offer better security for credit and loans.

The tables below combine to give a concentrated picture of industrial enterprise in workers' settlements, the size of plants measured in terms of output and the output of the main branches of industry.

Number of Workers in Main Industrial Branches

Branch	Number	Per cent
Timber	1,666	31·7
Metal	1,292	24·6
Food and tobacco	1,128	21·5
Glass and ceramics	230	4·4
Printing and paper	218	4·1
Chemicals	148	2·8
Leather	144	2·7
Miscellaneous*	431	8·2
Total	5,257	100·0

* Electric goods, cement and products, quarries, rubber goods and plastics, textiles, etc. Each of these branches accounts for 1–2 per cent of the total number of workers.

Size of Industrial Enterprises in Workers' Settlements
(in terms of number of workers and gross output)

No. of workers	No. of plants	No. of workers employed	Output (IL. millions)	Per cent		
				Plants	Workers	Output
−10 workers	45	300	4·6	30·8	5·7	5·4
11–20	38	584	7·1	26·0	11·1	8·4
21–30	24	608	9·3	16·4	11·6	11·0
31–40	12	426	7·5	8·2	8·1	8·9
41–50	4	191	3·6	2·8	3·6	4·3
51–75	11	719	9·7	7·5	13·7	11·6
76–100	4	355	7·8	2·7	6·8	9·4
101–150	3	365	7·8	2·1	7·0	9·2
151–300	3	608	9·0	2·1	11·6	10·7
301–	2	1,095	17·7	1·4	20·8	21·1
	146	5,251	84·1	100·0	100·0	100·0

The tables above and overleaf show:

(1) The timber, metal and food industries employ more than three-quarters of the workers. They account for a similar proportion of the gross output. One of the reasons may be that the main products of the timber industry—builders' joinery and furniture—are absorbed by the agrieultural settlements themselves. Plywood as stated, of

Size of Plants (in terms of output and employment), 1959

Range of output	No. of plants	No. of workers employed	Output (IL. millions)	Per cent		
				Plants	Workers	Output
IL. – 99,999	35	246	2·2	24·0	4·7	2·6
100,000– 249,999	43	620	7·1	29·5	11·8	8·5
250,000– 499,999	31	770	10·7	21·2	14·7	12·8
500,000– 999,999	21	994	15·4	14·4	18·9	18·4
1,000,000–1,499,999	4	262	4·5	2·7	5·0	5·3
1,500,000–1,999,999	3	355	5·2	2·0	6·8	6·1
2,000,000–2,499,999	3	332	6·8	2·0	6·3	8·1
2,500,000–3,499,999	2	192	6·1	1·4	3·7	7·2
3,500,000–4,499,999	2	385	8·4	1·4	7·3	10·0
5,000,000–	2	1,095	17·7	1·4	20·8	21·0
Total	146	5,251	84·1	100·0	100·0	100·0

Gross Output of Industry in Workers' Settlements, 1959

Branch	IL. millions	Per cent
Timber	26·7	31·8
Food and tobacco	21·9	26·1
Metal	17·2	20·5
Chemicals	4·6	5·4
Rubber and plastics	2·5	3·0
Glass and ceramics	2·3	2·7
Printing and paper	1·8	2·1
Miscellaneous *	7·1	8·4
	84·1	100·0

* Electrical goods, cement and products, quarries, leather, textiles, etc., each branch accounting for under 2 per cent of total output.

course, is produced mainly for export. This is also the case with the output of the metal industry, which is made up largely of agricultural machinery and equipment, irrigation fittings and accessories and the like. Indeed, these timber and metal factories constitute an advanced stage of the workshops established in every *kibbutz* to meet local requirements; they were not planned and set up originally to serve an outside market. The food industries exist mainly for the processing of locally available agricultural raw materials.

(2) The development of other branches of industry, still in their

initial stages of development, such as textiles or leather goods, and perhaps also cement, quarries, paper, rubber, electrical goods, precision instruments, electronics and optical goods, fine crafts, etc., may also be feasible. The manifold opportunities offered by the processing and storage of diverse products of the farm have not yet been fully exploited.

(3) Most of the plants—again about three-quarters—are small, employing less than thirty workers. By contemporary standards they are hardly more than workshops. Factories in the currently accepted sense of the term are few.

Sixty-nine per cent of those employed in industry in the workers' settlements are concentrated in 23 larger establishments with 50–500 employees or an average of 133 per plant. Thus it may be said, 23 factories—i.e. 15 per cent of the total worthy of the name (in terms of number of employees and value of production)—account for more than two-thirds of the output of this section of industry.

(4) Only 16 plants had an annual output exceeding IL. 1 million. The annual output of another 109 factories (again three-quarters of the total) was under IL. 500,000 each, while one-quarter had an annual output of less than IL. 100,000.

This provisional summary would seem to indicate that taking into account the economics of scale the overwhelming majority of the smaller enterprises cannot prove economically sound. And seeing that the need for non-agricultural enterprises in the settlements must increase, the apparent solution is the establishment of regional enterprises.

We shall return, after our chapter on municipalization, to this question of regional enterprise and the joint operation of factories by many agricultural settlements, not included within district municipal authorities.[1]

[1] Sources: (1) Statistical Abstracts and Industry in the Labour Economy (facts and figures) 1959-60.
(2) D. Shriro. 'Industry in Settlements and its Development'. *The Economic Quarterly* (Hebrew) No. 16, 1957, p. 395.

VII

MUNICIPALIZATION

SINCE the creation of the State of Israel a rapid process of municipalization has set in due to the tripling and wider distribution of the country's population, the establishment of new settlements and the expansion of existing ones. Local councils expanded and were granted the status of town councils. Municipal status was also given to localities which had hitherto lacked it and also to new settlements. A characteristic feature of this period was the setting up of regional councils. The following table sums up the dynamic expansion of municipal administration in Israel.[1]

Category of Local Authority	Year 1949	Year 1961
Total	34	184
(1) Municipal Councils	8	25
(2) Local Councils	20	109
(3) Village Councils	2	—
(4) Regional Councils	4	50

Out of a total population at the close of 1959, of 2,089,000 inhabitants, no more than 119,000 lived in settlements without any local governmental status. Thus 94·3 per cent of the inhabitants of the country came within the jurisdiction of municipal authorities. The growth of population within recognized municipal areas by far exceeded that of the general increase in population. The process of municipalization embraced not only the new immigrants, but also a large part of the established residents, including the Arab towns and villages.

[1] Statistics have been taken from the Ministry of Interior, Municipal Research Bureau; Local Authority Budgets (mimeographed) No. 4, 1961; also, 'The Finances of the Local Authorities, 1960'.

The number of people living in areas devoid of municipal status is steadily declining, as more and more settlements are swept up in the process of municipalization.

Fifteen local councils, mainly rural *moshavot* based upon private farming, have been raised to the status of town councils. New development towns have been recognized as local councils. Village council status has, for all practical purposes, been abolished. Villages which indicated their desire in that direction were granted the status of local councils. The latter are ordinary villages engaging in private farming, and do not include either *kibbutzim* or *moshavim*. Most are affiliated to regional councils or are in the process of affiliation.

Changing municipal status is a common phenomenon throughout the world. In the United Kingdom, for example, a special commission has been entrusted with the task of altering the administrative and municipal map. England has to cope with a peculiar problem in settling the bounds of jurisdiction of city councils and county councils. 'The system of local government in England reflects a kind of balance between the often divergent interests of counties and big towns.' The London *Economist* recently wrote on this question:[1]

'The present Local Government Commission, which is slowly redrawing the administrative map in England, is not authorized to change this system; but it does have considerable discretion to adapt it to modern requirements' . . .

. . . 'The dichotomy between county councils and town councils is both the strength and the weakness of the whole system. It is another of those curious English muddles that have somehow worked quite well. French logic has subordinated the local COMMUNE (however large) to prefectorial control exercised through the wider *departement*. American pragmatism has allowed the growth of cities and municipalities to reduce the counties to ineffectiveness. But in Britain county and "county borough" (i.e. big town) councils remain equally important, and each possesses a distinctive outlook. Few counties are really very rural today, yet a sort of gentlemanly rural atmosphere still pervades their activities' . . .

Some resemblance exists between the situation described by the *Economist* and the changing character of the fifteen large *moshavot* in Israel whose local governmental status previously was that of local councils and which have now become towns. The typical feature of the process of municipalization in Israel is the dynamic emergence and multiplication of the regional councils.

[1] City versus County. October 7, 1961, pp. 18, 19.

The fifty regional councils take in a population of 227,000 or about 11·5 per cent of the inhabitants of the country who are included in the framework of local government. In point of number of inhabitants the regional councils can be listed according to their districts as follows: Northern District (Jezreel, Acre, Kinneret and Safad sub-districts)—74,000; Central (Petah Tikvah, Rehovot, Sharon and Ramleh)—71,000; Southern (Beersheba and Ascalon)—47,000; Haifa (Haifa and Hadera sub-districts)—20,000; and Jerusalem District—15,000.

The largest sub-districts from the point of view of regional council organization are Jezreel with 32,000; Ascalon—27,000; and Sharon—28,000. In other sub-districts the regional councils account for less than 20,000 inhabitants. It is interesting that in the vicinity of the larger towns (excluding the Jerusalem Sub-District) there are very few regional councils. In the Haifa Sub-District they take in only 6,000 inhabitants and in the Tel Aviv District, comprising one-third of the total population of the country, there is only one regional council—Ono—comprising only four settlements. The separate existence of this council is unjustified and it should be amalgamated with the neighbouring council or abolished entirely.

The proportion of the population under municipal government which is included in regional councils near the other larger towns constitute 8 per cent in the Jerusalem District and 6 per cent in the Haifa District. In the Central District the proportion is larger—18 per cent, in the Northern District—29 per cent, and in the Southern District—33·5 per cent. The idea we are developing in the present study is based mainly on two Districts—the Northern and Southern—comprising 20 per cent of the county's inhabitants, and also certain sections of the Central District, the Hadera Sub-District and the Jerusalem District. We shall not discuss the Tel Aviv District and the environs of Jerusalem. Some distance from Haifa, in the direction of the Valley of Jezreel, lie other areas to which our plans apply.

The fifty regional councils vary widely in point of size, the number of settlements they comprise, their areas of jurisdiction and population. This is a new development, and undoubtedly, for many local reasons, there has been excessive fragmentation. At some time in the future, perhaps, councils will be merged to form larger units. Probably one-third of the present number would be adequate as there is little logic in having regional councils comprising no more than three or even six settlements. Some idea of the present state of affairs can be obtained from the following figures: 6 councils comprise up to 5 settlements; 18 councils, 6–10; 21, 11–20; and 5 more than 20. More than half of the regional councils comprise from 11 to 38 settlements each. The municipal authorities, and particularly the settlements and local leaders, must endeavour to unite to form larger units. To put the *agrindus* idea into effect larger regional units are a desideratum, the optimal number being over 20, for obviously size carries with it more favourable prospects for the development and maintenance of industrial centres. What we propose requires a concentration of strength and not fragmentation, as we shall explain in what follows.

The Israel Ministry of Interior has to ensure that the municipal authorities carry out their local governmental functions in accordance with the law. With this object in view it extends every possible assistance to them, though proper precautions are taken to prevent them from exceeding their authority. The municipal councils for their part do not normally try to encroach upon functions and spheres which are not properly theirs. This is also the case with the local councils. The regional councils, however, have seen their function in a different light. From the very outset most of them have ventured into economic enterprise. For example: they regard the operation of a large tractor or excavator as a 'service', as the individual settlement, be it even a large *kibbutz* or *moshav*, cannot purchase and maintain such expensive equipment. A slaughterhouse in an agricultural area differs from the urban abattoirs as its functions are more ramified and varied. It must make provision for cold storage and freezing, preparation, packing and even marketing. The rural slaughterhouse is an economic enterprise not a municipal service. The same applies to enterprises such as the curing and packing house and the cotton gin. These are the natural functions of regional association. The progress of agriculture, bountiful crops and occasional gluts, have accentuated problems such as storage, refrigeration, preparation, processing and packing. These functions, which are of a markedly industrial character, are beyond the capacity of individual settlements. They cannot always be assumed by central marketing agencies even if these are of a co-operative character. Every district is interested in undertaking at least some of these functions locally. For this reason any primary regional association, even if only for municipal purposes, must tackle the urgent economic problems of its area as well.

Moreover, the process of municipalization which gathered momentum after the creation of the State of Israel found an established tradition dating from the Mandatory period. Under the British administration the district had made representations to the authorities informally as a *gush*—bloc of settlements—in matters affecting security, roads, water, drainage, grazing rights, soil erosion, and various services. The *gush* would also handle questions like border disputes and land problems. The *gush* committee served as an unofficial regional authority. Actually the District Commissioner appointed a *mukhtar*—headman—for each village, but the most respected *mukhtar* of the district was assisted by a committee made up of representatives of the settlements in the district. The competence of this committee was much wider than that officially enjoyed by the *mukhtar*.

Under the Turkish régime the *mukhtar* acted in liaison between his village and the authorities. It was customary for the District Governor to appoint the incumbent to this office, and indeed until 1943 there was no law setting out the functions and authority of the headmen. In effect they were responsible for the maintenance of law and order in their settlements, were required to inform the nearest police station of any serious crime or incidents out of the ordinary, to assist government officials in the collection of taxes and other imposts, to publish all official notices and announcements despatched to them by the District Governors, to keep a registry of births and deaths—a copy of which was to be sent to the Chief Health Officer every three months. The *mukhtar* had also to safeguard the roads, railways, telephone and telegraph services, forests, antiquities and historic sites, and to inform the authorities of the outbreak of epidemics and pests. He was in a sense the local representative of the Government. The office persisted until quite recently even in areas in which properly constituted municipal authorities—not excluding the large towns—were in operation. The *gush* committees, which had been set up without any official status or sanction during the Mandatory administration, were administered by the leading *mukhtars* of the area or even without them. From these *gush* committees the regional municipal authorities developed.

The legal sanction formulated by agreement between the Ministry of Interior and the regional councils stipulated that the latter confine themselves exclusively to their municipal functions. For the maintenance of economic functions they were permitted to establish an economic instrument, which is registered as a society or company by the Ministry. In fact the same person can serve at one and the same time as Chairman of the Council and Chairman of such a society or company. Financial operations, however, are kept strictly separate. The society or company (as the case may be) is kept separate and comes under the supervision of the Registrar of Co-operative Societies or the Registrar of Companies. The ramification of economic operations led to an increase in the number of companies and societies, and enterprises managed independently by the regional councils or in partnership with other bodies—local or external, incorporated or private—multiplied.

The Ministry of Interior scrupulously distinguishes between services which normally came within the scope of the regional councils, and economic enterprises operated by a registered co-operative society. The declared intention being to concentrate all operations, within and about the Council, the Ministry has little fear of the possibility of confusion in spheres of competence. The Ministry has agreed that members of the Council also serve as directors of the co-operative society. It insists only—as already indicated—on the keeping of separate accounts and records.

Every regional council has its agricultural committee which may institute and maintain agricultural services such as disease and pest control.

The services of a more economic character—such as cold storage plants, cotton gins, packing houses and bakeries—must be operated as co-operative societies. Such societies, of course, being subject to liquidation and requiring credit, must come under the supervision of the Ministry of Interior and the Auditor-General.

The regional councils regard their economic functions virtually as their *raison d'être*, for they are undertaking operations which are beyond the capacity of single, even large, settlements. Indeed the latter continue to maintain a number of the more important services such as education, sanitation, the maintenance of security and public order. The Councils, however, also carry out the de-stoning of land, drainage and even the cultivation of experimental orange groves.

The following clauses from the Statutes for economic activities of the Shaar Hanegev Regional Council[1] are instructive:

Organizational Activities

(1) A holding company for all enterprises of the Council, in which all settlements shall hold an equal share, shall be set up.

(2) Every settlement shall have one vote. The meeting of all voters shall constitute the plenary session of the company and shall constitute the supreme authority in all joint activities.

(3) Every subsidiary enterprise of this company shall be managed independently, for economic, financial and vocational purposes. No enterprise shall transfer its profits to any other enterprise.

(4) *a.* Every settlement shall be a partner in all enterprises.

b. No enterprise shall be set up other than by the consent of 43 voters.

c. Any settlement compelled to discontinue the supply of raw materials to any enterprise shall be responsible for any losses incurred, unless it can prove that it itself suffered loss,

---

[1] Furnished by the Shaar Hanegev Regional Council.

and the settlement shall be exempt from the payment of compensation by a decision of the society in plenary session.

(6) The directors of the society shall serve as the management of the enterprises.

(8) The society in plenary session shall appoint a manager for economic operations who shall be responsible for research, planning, control of the enterprises, the organization of services, the raising of standards of efficiency and the development of new enterprises.

These statutes also make provision for the composition of boards of directors and the committees, admission of new members and the like. A highly detailed section governs the manner in which the constituent settlements shall participate in the directorate of the society, imposing upon them the obligation to second suitable members to fill various offices in the district.

The organization of economic activity in the various regional councils is not uniform. Some councils have floated limited companies, others have formed co-operative societies. Yet others have established a society of more general character, engaging in all operations instituted by the council and the settlements. A number have set up a separate body to deal with each enterprise or field of activity. Generally the councils themselves are responsible for the supply of water, but there are also areas in which there is a separate water co-operative. This is particularly the case in areas where the latter was founded before the council. Councils of larger, highly developed agricultural regions tend to concentrate upon farm produce processing plants. In less advanced districts pride of place is given to services, especially those catering for consumers, e.g. regional bakeries.

In all these matters the policy of the Ministry of Interior is shaped in close consultation with the regional councils both directly and through national conferences held to discuss different problems, both municipal and economic, which are among the chief concerns of the Ministry and council chairmen.

The Chairman of one regional council, replying to a question put by the Ministry as to why he had floated a limited company for a certain enterprise instead of setting up a co-operative society—seeing that all the constituent settlements of the district were registered as co-operatives—stated:

'This form (viz. a limited company) also allows for full public supervision . . . The main reasons for choosing the form of a limited company are: (a) We have taken into account the prospect of col-

66

laboration of outside capital in the enterprises upon the basis of partnership. This can be secured only by a limited company. (*b*) . . . The settlements possess the advantage that in the event of other bodies joining the company they will hold the ordinary shares and will have the right to manage the company.'

Thus there is a considerable number of companies though most enterprises of this character have been registered as co-operative societies.

Another answer to a Ministry of Interior query went as follows:

'A co-operative society in which the eighteen agricultural settlements in the area are members has been set up . . . The council is not a member of the society, but the chairman of the council is also the chairman of the society, the executive of the council is the executive of the society and the plenary session of the council is the general meeting of the society' . . .

The chairman of another regional council replied to a similar question in the following terms:

'We have two economic enterprises—one for heavy agricultural equipment and a regional citrus packing house. Our council has resolved that both must be registered separately as co-operative societies. The chairman of the regional council is the chairman of the society for heavy agricultural equipment.

. . . The interests of the council in the society will be safeguarded as a shareholder.

. . . The council shares no liability for any losses incurred by the undertaking.'

We have before us the balance sheets of the regional councils, as well as their budgets prepared by the Ministry of Interior, which obviously exercises close supervision of the financial operations of all municipal authorities, besides drafting Annual Surveys of the authorized budgets of local governmental bodies complete with a general report and detailed tables on the operations of each of them.

The Survey for 1960/61 covers 161 authorities; 12 are not included.

Our interest is confined to the regional councils, but we shall cite certain more general data to enable us to assess the part they play in purely local government affairs within the wider municipal context.

Estimated Ordinary Revenue of Municipal Bodies
(in keeping with local governmental status)—1960/61

Total no. of local authorities	216
of which: Municipal Councils	167
Local Councils	31
Regional Councils	18

67

The estimated revenue of the regional councils in this year came from the following sources:

Rates	IL. 2·9 million	16	per cent
Local services	1·3	7·4	
National services	1·7	9·2	
Enterprises	2·5	14·2	
Government grants	5·5	30·7	
Owners' participations	3·7	20·5	
Loans and sundries	0·6	2·0	
Total	IL. 18·2 million	100·0 per cent	

Owners' participations include the instalments paid by local committees on account of repayment of loans.

This item accounts for 20 per cent of the estimated revenue of the regional councils but only 3–4 per cent of that of the local and municipal councils. But if we take it together with rates the proportion in both the former and the latter is about equal.

The Survey, noting increasing revenue from municipal rates, attributes this growth to the following causes: the rise in the number of inhabitants, the increase in the number of buildings and businesses, higher rates and changes in local methods of taxation, more effective tax collection and higher price levels.

In the course of the past two years, indeed, the national average increase in municipal revenue has been 42·4 per cent. The regional councils averaged 40·2 per cent. The figure for the municipal councils was lower than the national average, while that for the local councils was higher.

Over a period of four years revenue per resident from municipal rates has risen from IL. 34 to IL. 53, that is by 56 per cent. In this same period the Government's taxation revenue per resident rose by 70 per cent. Thus the growth in revenue from municipal rates lagged considerably behind the increase in Government taxation.

Government grants-in-aid help cover the costs of education and culture, health, social welfare, relief works and general purposes. It is interesting to note that the increase in Government grants to municipal bodies has been lowest in respect of regional councils.

General administration	9·4
Local services	31·1
State services	34·0
Enterprises	15·4
Repayment of loans for consolidation	7·8
Miscellaneous	2·3
Total	100·0

Over the past two years Government grants to regional councils grew by 17 per cent as compared with an increase of 39 per cent to municipal councils and 70 per cent to local councils.

A breakdown of the expenditure of the regional councils for 1960/61 is given on previous page (in percentages).

In the regional councils the item 'Repayment of loans for consolidation' is of a specific character, differing from the other municipal authorities as it also includes sums paid on account of certain economic undertakings of the various settlements.

Local services include sanitary services, local security, planning and construction, maintenance of municipal property, agricultural services and miscellaneous. State services include education and culture, health, social welfare, religious services (exclusive of Government grants to the religious councils).

Income from services covers only a small proportion of the expenditure (18 per cent of local services and 22 per cent of State services). The difference is made up by rates and other revenue. The Government's grant covers 18 per cent of the cost of State services. Thus the local authorities must cover 60 per cent of the cost of these services by rates and other revenue. It must be noted, however, that the expenditure on education does not include the salaries of school and kindergarten teachers which are paid direct by the Government.

The 50 regional councils differ widely in respect of the size of their budgets. In 5 councils they run into five figures, in 3 to seven and in 42 to six. The budget of the largest council aggregates more than IL. 2 million, followed by a council with a budget of IL. 1·7 million and a third with a budget of over one million. After this 'big three' come five councils with budgets exceeding IL. 500,000, 22 in the IL. 250,000–500,000 range and 15 in the IL. 100,000–250,000 range. There are five small councils whose budgets are commensurate with their size.

A more detailed analysis of the budget of one regional council should prove highly instructive in this respect. We shall choose, for our purpose, a council with a medium-sized budget, not among the first three whose annual expenditure ranges between one and two millions, but the fourth on the list, whose authorized budget for the year 1960/61 was IL. 625,000. For this council we have also the estimates for the current financial year—1961/62—totalling IL. 896,000. This example can serve to illustrate the situation and activities of other regional councils.

The local services include sanitation, local security, planning and construction, municipal property, agricultural services, immigrants, settlements, etc. The State services are education, culture, health,

MUNICIPALIZATION

Revenue	Authorized Budget 1960/61 (IL.)	Estimates 1961/62 (IL.)
Taxation		
Rates	70,540	79,250
Licences, Fees, etc.	17,050	31,050
Total, Taxation	87,590	110,300
Income from Local Services		
Sanitary Services	5,000	6,500
Local Security	25,400	185,290
Planning and Construction	4,500	9,400
Revenue from Council Property	6,800	10,300
Agricultural Services	50,000	79,000
Miscellaneous	14,000	10,500
Total, Income from Local Services	105,700	300,990
Income from State Services		
Education	20,600	27,400
Culture	20,970	27,250
Social Welfare	—	200
Religious Services	4,590	5,160
Total, Income from State Services	46,160	60,010
Enterprises		
Amphitheatre	45,000	55,000
Government grants-in-aid		
Education	24,500	27,000
Social Welfare	6,000	7,000
General purposes grant	58,000	67,000
Miscellaneous		
Participation in Religious Council Budget	1,800	2,560
Total, Government grants-in-aid	90,300	103,560
Owners' participations	219,363	265,570
Total ordinary income	594,113	895,430
Estimated surplus 1.4.61–31.3.62	30,658	239
Total, Income	624,771	895,669

Revenue	Authorized Budget 1960/61 (IL.)	Estimates 1961/62 (IL.)
Estimated Expenditure		
General administration	49,600	56,370
Local services	276,278	471,480
State services	125,570	148,450
Enterprises	50,400	103,670
Repayment of loans and consolidation	116,431	112,040
Estimated surplus	6,492	3,659
Total	624,771	895,669

social welfare, religious services. The economic enterprise of the regional council is the district amphitheatre.

The example we have taken is the budget of a larger regional council (though, as stated, it is not among the largest) representing 18 *kibbutzim*, some of them among the oldest in the country, situated in a richly endowed and intensively cultivated area. The budget, however, only reflects municipal activity and only tells half the story. It does not include drainage work, in which over a period of years the council has invested hundreds of thousands of pounds. It tells nothing of the plants processing agricultural produce, of the packing houses, the large creamery, and particularly the factory, which is one of the most important of its kind in the country.

We shall discuss all these in a special chapter on regional enterprises as the purpose of the present section is to describe one specific function of the regional council—its municipal function—which in itself, needless to say, is of major importance.

It is not irrelevant to state here that in the view of the author an excessively large number of regional councils have been set up and a number of the smaller bodies should be merged with larger neighbouring councils.

Map 1 shows the boundaries of all existing councils as well as the number of settlements each includes. Map 2 indicates boundary changes considered desirable in areas suitable for the establishment of *agrindus*.

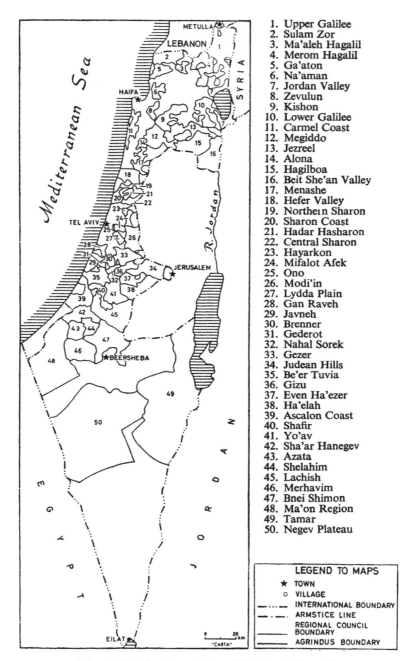

1. Upper Galilee
2. Sulam Zor
3. Ma'aleh Hagalil
4. Merom Hagalil
5. Ga'aton
6. Na'aman
7. Jordan Valley
8. Zevulun
9. Kishon
10. Lower Galilee
11. Carmel Coast
12. Megiddo
13. Jezreel
14. Alona
15. Hagilboa
16. Beit She'an Valley
17. Menashe
18. Hefer Valley
19. Northern Sharon
20. Sharon Coast
21. Hadar Hasharon
22. Central Sharon
23. Hayarkon
24. Mifalot Afek
25. Ono
26. Modi'in
27. Lydda Plain
28. Gan Raveh
29. Javneh
30. Brenner
31. Gederot
32. Nahal Sorek
33. Gezer
34. Judean Hills
35. Be'er Tuvia
36. Gizu
37. Even Ha'ezer
38. Ha'elah
39. Ascalon Coast
40. Shafir
41. Yo'av
42. Sha'ar Hanegev
43. Azata
44. Shelahim
45. Lachish
46. Merhavim
47. Bnei Shimon
48. Ma'on Region
49. Tamar
50. Negev Plateau

LEGEND TO MAPS

★ TOWN
○ VILLAGE
-·-··- INTERNATIONAL BOUNDARY
-·-·- ARMSTICE LINE
—— REGIONAL COUNCIL BOUNDARY
═══ AGRINDUS BOUNDARY

Map 1. Existing Boundaries of Regional Councils

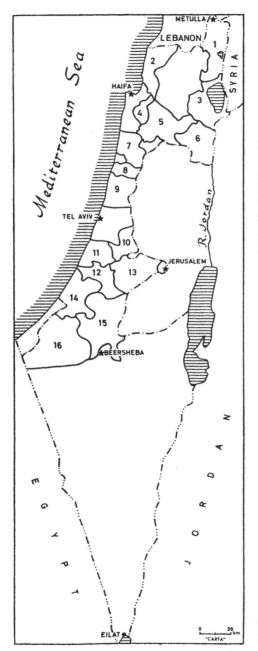

1. Upper Galilee
2. Sulam Zor, Ga'aton, Na'aman
3. Jordan Valley, Lower Galilee
4. Zevulun
5. Kishon, Megiddo, Jezreel
6. Beit She'an, Gilboa
7. Alona, Menashe
8. Hefer Valley
9. Sharon (Consolidated)
10. Javneh, Gederot, Nahal Sorek, Brenner, Gan Raveh
11. Gezer, Modi'in
12. Be'er Tuvia, Yo'av
13. He'elah, Judean Hills, Gizu, Even Ha'ezer
14. Ascalon Coast, Sha'ar Hanegev
15. Lachish, Bnei Shimon
16. Ma'on Region, B'sor Region

Map 2. Projected Boundaries of Regional Councils

73

VIII

INTER-REGIONAL ENTERPRISE

A T the beginning of the Second World War when a system of food rationing and control was introduced, the agricultural settlements were apprehensive that Hamashbir Hamercazi, the central supply institution, might not be able to carry out its functions as effectively as before because of the new conditions, over which, of course, it had no control. One of the results was the association of groups of settlements in making joint purchases, which they were able to do at lower prices and better conditions of credit. Structurally these associations were based upon geographical proximity and grouping-affiliation. In this way the 'buying organization' was set up by seven neighbouring villages in Hefer Valley and Samaria. The organization proved successful and other settlements began to join. After twenty years it comprised thirty-eight *kibbutzim*, situated between Haifa and Tel Aviv. Today it is called 'The Hefer Valley and Samaria Settlements Organization', but it goes beyond these two areas, embracing six regional councils. It is no longer confined to a single movement, and the associated settlements belong to four *kibbutz* movements, while one is 'unaffiliated'.

The associated settlements occupy 180,000 dunams[1] of land (45,000 acres). They have 75,000 dunams (about 19,000 acres) under irrigation, 22,500 dunams (5,600 acres) of which are under groves, orchards and other plantations.[2]

Their annual production includes: 10 million litres of cows' milk; 300,000 litres of sheep's milk; 40 million eggs; 2,500 tons of dressed poultry; 750 tons of beef; 10,000 tons of potatoes and vegetables (in equal proportions) and about 300,000 crates of citrus for export.

The value of farm output marketed was of a magnitude of IL. 22 million.

[1] 1 dunam, metric = 1,000 m² = 0·247 acre.
[2] For further information see: 'The Hefer Valley and Samaria Settlements 1940–60', a report submitted to the Twelfth Council, 1961.

74

The share capital of the Organization is today IL. 100,000 and its balance sheet totalled IL. 5·3 million. Its liabilities were as follows:

Banks	IL. 1·3 million	26 per cent
Hamashbir Hamercazi (open account)	1·3	26
Suppliers (open account)	2·5	48
	IL. 5·1 million	100 per cent

During the past year the purchases of the Organization totalled IL. 16 million (IL. 1·25 million per month). Thus the ratio of monthly purchases to debt was 1 : 4.

In 1960 normal purchases were made up as follows: provisions—20 per cent; cattle and poultry feed—41 per cent; farm supplies (seeds, chemicals, fertilizers)—15 per cent; fuel—14 per cent; sundries—10 per cent. The Organization has its own filling station in Hefer Valley operated by the Hefer Valley Co-operative.

One of the chief purposes of the Organization was to secure suitable terms of credit for its purchases.

Conditions have been laid down for the granting of credit to associated settlements based upon seniority, the number of inhabitants at the end of the previous year, and the overall income of the settlement as reflected in the last audited balance sheet. In addition to the conventional categories of credit—fixed, short-term purchases on credit and open account—many and various financial arrangements have been made, mainly for transfers between institutions with which the settlements are connected. The Organization has imposed no ceiling upon the purchases of the individual settlements, though it has limits for the various categories of credit. Special rules for security and mutual guarantees have been instituted.

Multiplication of such buying organizations facing similar problems has led to the establishment of a permanent secretariat to determine conditions of buying and financing purchases. The secretariat negotiates with the central bank (the Bank of Israel) and the Ministry of Agriculture about the volume of credit, which is governed by the variable dimensions of agricultural output, the volume of purchases and means of payment. Then the executive of the secretariat conducts joint talks with the banks on the allocation of credits to the organizations.

The importance of the secretariat is steadily being enhanced, as the organizations have learnt from experience that concerted action obviates independent effort being made on the part of one organization or another to secure suitable conditions for the self-same products.

Purchasing is the principal function of the Organization, but if

75

this were its sole function we should not discuss it in the present context. It is of special interest for us because, more recently, it has engaged in the establishment of joint projects.

In 1953 the Ministry of Agriculture proposed that the Hefer Valley region establish one of the projected lucerne meal mills. I recall that at the time machinery and equipment for three such mills had been brought to Israel; they remained, however, for eighteen months in the customs sheds before the Ministry succeeded in persuading three regions to instal and operate them. The *kibbutzim* of Hefer Valley agreed to enter into a partnership to operate the mill, but the *moshavim* in the area were reluctant to do so. The plant which, like the two others elsewhere, was, so to speak, imposed upon this region, once it had proved itself persuaded the settlements there to think of other enterprises. Interestingly enough, the two other mills furnished a similar stimulus in the areas in which they were installed.

The Organization now began to look around for a suitable location for its industrial enterprises. After much discussion a site east of the Hadera railway station was fixed upon. The lucerne mill was reinstalled here and commenced operation again in the spring of 1960. The mill has been operated for six seasons, steadily developing until it now dries and grinds 14,500 tons of lucerne (grown on an area of 3,700 dunams—925 acres) to produce about 4,000 tons of meal. The ratio of green lucerne to meal is thus approximately 4 : 1. Each year the plant is extending the operational season which now stands at 222 days in the year. The value of output in relation to every pound in the balance sheet is IL. 1·74, and to every pound of means of production is IL. 2·16.

In addition to grinding lucerne this project has undertaken other operations, including the production of fodder, seed-growing, spraying (both surface and aerial) the lucerne fields cultivated for drying, and the application of liquid fertilizer.

A feed mixing plant was the second enterprise to be set up by this Organization. This project was suggested to the Hefer Valley and Samaria settlements in the early fifties by the management of the Israel Agricultural Bank and an engineer who was an expert on mills and silo-elevators. Years passed; the directors of the Organization were changed several times before a planning committee was appointed in 1956 and before the work of construction began in 1960.

The Ambar feed mixing enterprise comprises: barns equipped with thirty-four compartments for different varieties of grain, meals, bran, oil-cake, milled orange peel, carobs, sugar beet-pulp, etc.; plant for peeling, pressing, milling, mixing and preparing pellets; warehouses, stores and auxiliary installations (a laboratory, metal workshop, as well as services for all the enterprises).

76

The managers of the central mill quote the price of feed to underline its economic advantages. The cost of feed is IL. 205·55 per ton at the mill, IL. 217·15 in a super-marginal settlement and IL. 219·80 in an ordinary settlement. Even in respect of a super-marginal settlement with an annual output of 60,000 tons the saving is sufficient to cover the initial investment in the mill within a period of four years. Further economy is secured by preparing the feed in high quality, clean pellets exactly weighed and mixed in the correct proportions, enabling maximum utilization. All these factors taken together exercise a very beneficial effect both on the health of the livestock and the profitability of the enterprise.

At the time of writing the central mill is already in operation; only a week ago we saw the first pellets to be produced. There are still a large number of problems of organization which the enterprise has to solve. It is managed by a secretariat elected by its workers who are advised on the composition of the feed by committees of specialists for the various branches of agriculture. A poultry expert is also employed. A settlement in the vicinity will serve for experiments with different varieties of feed, and two agricultural experiment stations have promised to co-operate. The question of how the feed is to be marketed (either through the Organization or through the mill as an independent economic unit) has yet to be settled.

KIRUR—COLD AND CURING STORES

Steady progress in the cultivation of bananas, apples and table grapes has prompted the Organization to seek an overall solution to the problem of cold storage. Bananas are picked while still green and have to go through a process of curing, which, until quite recently, was undertaken in the towns; in 1957, however, the Jordan Valley settlements set up the first regional curing house. A year later the Hefer Valley and Western Galilee settlements also recognized the advantage of local curing.

The bananas received at the store are cut and packed with the aid of special equipment before they are transferred to the curing rooms. The marketing of boxed bananas has led to increased demand, thereby relieving pressure on prices. Processing has also facilitated the marketing of second grade bananas in the neighbourhood; fruit that previously remained with the growers after the selection and despatch of the higher quality bananas to the central stores in the town is now also accepted. As a result of the introduction of careful grading methods, first-grade fruit now constitutes 55 per cent of the crop, while third grade has dropped to 10 per cent.

The curing houses have reached agreement with exporters and

now co-operate in the packing of fruit for shipment abroad. Exports are undertaken on the joint account of all growers in the country.

COLD STORAGE OF FRUIT

Apples

The existing department for apples has a capacity of 250 tons, but before it was fully complete work commenced on the planning of a new wing which can store another 650 tons, to meet the steadily growing needs of the Organization.

Grapes

A method of storage for table grapes has not yet emerged from the experimental stage, though the trial storage of 19 tons of fruit has already proved successful. Cold storage of grapes will enable growers to regulate consignments in keeping with marketing conditions.

MEAT AND OTHER PRODUCE

The cold storage plant is equipped with a deep-freeze room as well as three others for the storage of meat. The settlements of the Organization receive for their own consumption 12 tons of packed and unpacked meat every month. A two-months supply is kept on hand. The settlements also keep their fish-fillet and hard cheese supplies in the store.

The Organization's twenty years of bulk purchasing have proved highly advantageous for all of its members. A new and very significant phase in its development set in, following the acquisition of a 150-dunam (37½ acres) site, upon which it erected its lucerne mill, the Ambar feed mixing plant and a cold storage and curing house. Other enterprises are to be built on this site in the future.

The leaders of this association of 38 kibbutzim, situated between Haifa and Tel Aviv, have probably never entertained the idea that they are in a stage of transition, on the way to a new form of organization. It has already been pointed out that the settlements belong to 6 regional councils, which, however, comprise 94 settlements in all, moshavim as well as kibbutzim.

The Organization's 'industrial zone' is in the Menashe Region, of whose 18 kibbutzim only 8 are members. These 18 settlements could serve as the nucleus for the foundation of a Menashe Agrindus. In that case perhaps 150 dunams would prove insufficient for the urban centre of the agrindus, and its geographic location might be found unsuitable. Some other use would undoubtedly be found for it. Perhaps the existing services—the lucerne mill, the feed plant and the cold stores—would serve the 2 adjacent areas: the *agrindus* of Menashe counting 18 settlements and that of Carmel Coast with 19 (of which only 7 are members of the present Organization). 37 settlements would benefit from the union of these two *agrindi* for the joint operation of the three services, though the geographic grouping would then be different.

The Hefer Valley which comprises 38 settlements (only 10 *kibbutzim* belong to the Organization) could serve as an ideal framework for an *agrindus*. It offers excellent conditions for both town planning and for economic and social development. In the latter respect it is probably even superior to Upper Galilee, which, provided certain changes are introduced, can be regarded, *post factum*, as an *agrindus* already in existence. We shall, however, discuss this latter project in more detail below.

Central Sharon which has 9 settlements and Sharon Coast with 11 must combine and form an *agrindus* of 20 settlements (of which, incidentally, only 9 belong to the Hefer Valley and Samaria Settlements Organization).

Three *kibbutzim*, belonging to the *Mifalot Afek* regional council, which are members of the Organization, are situated 25 miles from the 'industrial zone' of the latter, near Hadera. They constitute an irrational appendix of the Organization and should join hands with the six agricultural villages in their neighbourhood. The one regional council which has only 3 settlements should, of course, be liquidated and attached to Mifalot Afek. The settlements of the Modi'in council (Renatia, Nahalim, Mazor, Nofech, Be'erot Yitzhak) should also be brought into this new grouping, thereby enabling the construction of a new and larger regional council containing 18–20 settlements which may, at some future date, form an *agrindus*. Why should three settlements of the area be affiliated to a distant organization, when some twenty concentrated in a single district could derive so much advantage from closer association, provided the necessary reorganization was introduced?

Map 3 is of the Organization of 38 *kibbutzim* and of other agricultural settlements which are not included. Map 4 indicates the boundaries of projected *agrindi* for part of the area.

Joint inter-regional economic and social undertakings are not to be excluded. The day is probably not far off when bigger enterprises will have to be set up to serve a larger number of settlements over a wider area, taking in several *agrindi* or regional councils. Three or four sugar refineries throughout the country should, we feel, belong to the beet-growers of the country. The first sugar-beet refinery was set up by the two central co-operative societies for marketing and

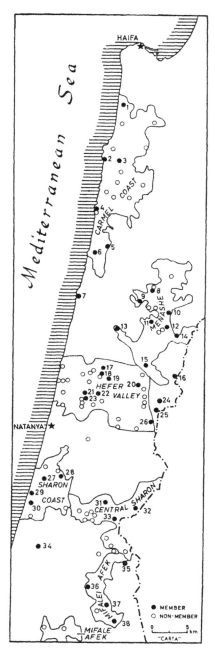

1. Hahotrim
2. Neve Yam
3. Ein Carmel
4. Nahsholim
5. Ma'ayan Zvi
6. Ma'agan Michael
7. S'dot Yam
8. Kfar Glickson
9. Mishmarot
10. Barkai
11. Ein Shemer
12. Ma'anit
13. Gan Shmuel
14. Metzer
15. Lehavot Haviva
16. Magal
17. Givat Haim (Ihud)
18. Givat Haim (Meuhad)
19. Ein Hahoresh
20. Hama'apil
21. Ma'abarot
22. Ha'ogen
23. Mishmar Hasharon
24. Bahan
25. Yad Hanna Senesh
26. Yad Hanna
27. Yakum
28. Tel Yitzhak
29. Ga'ash
30. Shefa'im
31. Ramat Hakovesh
32. Eyal
33. Nir Eliahu
34. Glil Yam
35. Horshim
36. Givat Hashlosha
37. Einat
38. Nahshonim

Map 3. Inter-regional
Organization

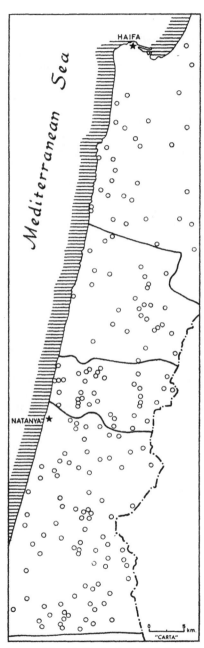

Map 4. Projected *agrindi*
81

supply (Tnuva and Hamashbir Hamercazi) and the Israel Bank of Agriculture. The second in Ramat Gan and the third in Kiryat Gat belong to private owners, but Government granted both loans amounting to about three-quarters of the capital invested. As the Ministry of Trade imports cheap sugar (65 per cent of consumption) it buys local sugar from the refineries at a price fixed and controlled by it. The main component in sugar production is beet. Governments are under constant pressure from sugar growers in the U.S.A. and other countries. From year to year growers increase or decrease beet acreage according to prices. It seems that it would be in the public interest if beet growers owned the refineries or were at least partners. We may cite the case of Holland where six sugar factories are privately owned, and another six co-operatively. Co-operative sugar factories owned by the growers could contribute largely towards strengthening ties between the prospective *agrindi* in the country and, until such *agrindi* are established, within and between the various regions. Our experience in similar cases in Israel has proved that strong co-operative enterprises such as Tnuva (in marketing) and Hamashbir (in supply) are to the advantage of the whole population, for consumers no less than the authorities controlling prices and marketing. Other enterprises of a similar character may be considered. There are already today undertakings based upon several regional councils, such as cotton gins and slaughterhouses (for both poultry and cattle).

Economies of scale will determine the establishment of such larger, joint enterprises, when the advantages of size are weighed against higher transport costs. It is, however, important that such partnerships should be set up by a number of *agrindi*, which again are based upon the principles of co-operation.

In total, it seems that all that these planners and executors lack is the proper perspective and that under certain conditions there is no basic contradiction between existing and future regional projects and the *agrindi* as planned.

IX

UPPER GALILEE

THE northern section of the Upper Galilean highlands lies outside the borders of Israel; the lower section, however, which is Israel's territory, includes some of the highest mountains in the country. These mountains are composed of hard dolomite. The mountain range, which begins in the neighbourhood of the city of Safad, is known as the Mountains of Naphtali. This range constitutes the main watershed of streams flowing both to the Mediterranean and to the Jordan and Lake Hulah. It is a typical highland region devoid of high plateaux or extensive plains, and without large basins or wide valleys. The largest and prettiest valley is the Vale of Kadesh Naphtali.

The many springs, found in the smaller basins, form streams which flow down the slopes, saturating the soil and producing a luxuriant vegetation. At one time the area was wooded. Today the unfavourable topographical formation constitutes a serious obstacle to cultivation. Despite the abundant rainfall, the inadequate layer of soil is unable to retain sufficient moisture to nourish summer crops, and irrigation is essential.

Even the thriving orchards in this area (mainly of apples) are in need of irrigation. And because of the unfavourable topographical conditions, only restricted areas of irrigable land suitable for fodder crops (which provide the basis for dairy-farming) are available. On the other hand, the rich pastures of these highlands are eminently suitable for sheep-raising, and the luxuriant and variegated flora provide a rich source of nectar for bees.

Agriculture in this area was revolutionized in the last few years, after the completion of a local irrigation project, leading water from the bountiful Malcha springs by a pipeline running from north to south, parallel to the main highway. Before long the hill settlements expect to obtain water for irrigation, though the excessively high cost will present a major problem for agriculture. A solution may be found if it should eventually be decided to fix a standard price throughout the country for water used for irrigation purposes. The Water Bill, 1959, establishes an Equalization Fund. The users of irrigation water in the regions where water is relatively cheap, consumers of drinking water in towns and villages and industrial enterprises pay progressive fees. The regions where water is scarce and expensive (in the south) are subsidized by the Equalization Fund.

THE HULAH VALLEY

The Hulah Valley is a remarkably level plain and for this reason, as well as because of the natural obstacles blocking the mouth of the lake, extensive marshes have been formed. Irrigation plans envisage the removal of these obstacles and the reclamation of a large part of the lake. Topographical conditions in most parts of the valley, with the exception of the settlements on the slopes along its perimeter, are extremely favourable. The soil here is mainly of alluvial clay. The land is fertile and all the conditions for the development of irrigated farming exist. Dairy-farming, based upon various types of cultivated fodder and particularly lucerne, flourishes, while vegetable growing and irrigated orchards—mainly apple—are other thriving branches of local agriculture.

Sheep-raising is based upon the natural pastures of the foothills, while good conditions for the development of artificial pastures indicate favourable prospects for breeding beef-cattle. Industrial crops, too, including cotton and ground-nuts, produce good results. Abundant supplies of water have stimulated the development of carp-breeding ponds.

Extensive, and often excessive, irrigation makes the improvement of the soil by drainage necessary.

The area totals 284,000 dunams (71,000) acres. In 1896 the first *moshava*, and in 1916 the first two *kibbutzim*, were established in this area.

A more intensive agricultural settlement began in 1939 after the redemption of the Huleh lake concession. By the time of the establishment of the State of Israel in 1948 the Jewish population of the region totalled 4,500 persons, living in eighteen villages, most of them *kibbutzim*.[1] By 1961 population has grown to more than 26,000.

This region, however, is considered to be still far from the saturation point in relation to the absorptive capacity of this region.

Of the 38 Upper Galilee settlements 33 belong to a regional council while 5 have local councils (2 of these are towns and 3 *moshavot*).

[1] H. Halperin, *Changing Patterns in Israel Agriculture*, Routledge & Kegan Paul, London, 1957, pp. 8-10.

They have 26,000 inhabitants in all, 16,000 of whom live in the towns and 10,000 in the villages.

The regional council of Upper Galilee was established in 1950 as a direct successor to the Gush Committee (of the type already referred to). At this time about 100,000 new immigrants were housed in hastily erected transit work camps in various parts of the country, impatiently awaiting more permanent employment and accommodation. The agricultural settlements and the regional council in which they were organized embarked upon a resolute effort to absorb the newcomers in their area. In the abandoned Arab town of Halsa and the veteran *moshava* of Rosh Pina work camps were established to provide, temporarily at least, for thousands of immigrants. Scores of settlers from the neighbouring *kibbutzim* volunteered to train the newcomers, to secure employment for them on the farms, to teach them Hebrew, to establish schools for their children, to safeguard their health and to insure minimum living conditions. In the course of the first year more than 7,000 immigrants were absorbed—more than the established population of the area.

The regional council undertook primary development works in Upper Galilee as part of its efforts to settle the immigrants: road-building, the construction of pavements, sewerage, waterworks, anti-malaria operations, local security, an ambulance and first aid service, slaughterhouse, mobile post office, social welfare services for immigrants, the construction of school buildings, cultural centres and an amphitheatre.

In the second year after the creation of this council the town of Kiryat Shmonah (formerly Halsa) was launched. The establishment of the new town was such a spontaneous act that the founding fathers are unable to recall how exactly it came into being. Marking its tenth anniversary at the beginning of 1960 one of the citizens wrote: 'No one can really reply to the question: How did it all begin in Kiryat Shmonah? The answer is simply: It just grew!' However, the veterans of the areas have a better memory than the new immigrants and they relate how the Head of the Jewish Agency's Settlement Department came from Jerusalem to inform them that the regional council's proposal to set up the urban settlement of Halsa had been approved. The council's leaders, taken aback by the announcement, asked apprehensively: 'How do we set about it?' To which the Jewish Agency representative replied: 'We'll give you a Swedish pre-fab. Move your office into it. The town will be built up round the hut!' That, in fact, is how the nucleus was formed. In the council's hut plans were drafted. But action preceded planning. And some years later the Minister of Labour, during a visit, declared: 'Of all the artificial towns in Israel, Kiryat Shmonah is the most

85

natural!' How external bodies conceived the aid they would extend was pithily summed up in a remark by an officer of the Agricultural Workers Union: 'Our job is to shove you up to the crest of the hill. Then we will give you a good kick and you'll roll down yourselves.'

Underlying these statements was the reluctance of central institutions to promise more than they could hope to fulfil. The local leaders, however, acted differently. They constantly urged the new settlers on but the latter scoffed at them and their plans. They looked upon the heads of the regional council as unpractical dreamers.

But in the last resort it was the regional council which drew up the plans, guided the settlers and set up the projects. It organized a crèche, kindergartens, a school and a kitchen for children. It concerned itself about employment, the development of allotments and the provision of permanent housing. It even opened a cinema, arranged for various performances from time to time, and constructed an amphitheatre for this purpose.

The need to provide the settlers with work prompted the flotation of various joint enterprises by the settlements. In the council's second year, already a regional nursery for fruit trees, a garage and a bakery were established. These were followed by an ice factory. Approach roads were laid down and a loan fund established, while the council set about organizing its technical department. In the fourth year an agricultural experiment station and a soil research laboratory were founded; machinery and equipment for a quarry were purchased; the exploitation of Hulah peat began; a local power station was put into operation; and ambulances and fire-engines acquired. After this came extensive drainage works and the construction of a fruit and vegetables packing house. Avenues of trees were planted in the town and a clinic was set up.

At the same time the council had to continue various essential services and to implement plans made before it came into existence. These activities included anti-malaria work, the draining of the Hulah marshes, soil conservation, settlement of water rights, the irrigation of the Hulah Valley, drainage, the exploitation of peat deposits, iron mining, etc.

ANTI-MALARIA WORK

A number of areas of Israel were infested with malaria in the past. In some of these the disease was eradicated. Upper Galilee had always been stricken and all members of the local Arab population suffered from it. Anti-malaria work in the district, undertaken by settlement and medical institutions, had continued unremittingly for decades, and was given a new stimulus in 1939 when a number of

Jewish settlements were founded. In spite of the very substantial progress made, Upper Galilee remained constantly vulnerable to the incursions of the anopheles mosquitoes from across the long borders, for in the neighbouring territories no anti-malaria work worthy of the name had been undertaken.

Comparison of the marshlands in Israel generally, is sufficient to point out the unfavourable conditions—in this respect—prevailing in Upper Galilee.[1]

The 144 springs in this area constitute 25 per cent of all springs in Israel.

The 27 *wadis* (watercourses) and canals are 11 per cent of the total number.

The 6 brooks are 30 per cent of all brooks.

The 63 pools and lakes are 26 per cent of the total.

Prior to the drainage works referred to, the area under water in Upper Galilee was 50 per cent of the total area under water in Israel.

Persistent anti-malaria work has given good results. The incidence of the disease dropped from 23 per cent in 1939 to 3 per cent in 1945, while more recently only one case per thousand inhabitants has been registered.

We have outlined this activity in some detail, as previously it came within the scope of the central settlement and health authorities, but from 1954 was taken over by the regional council.

HULAH DRAINAGE PROJECT [2]

For many decades settlement bodies in this country had studied the draining of the Hulah marshes in the northern area. Since before the outbreak of the First World War the concession rights had passed from one Arab owner to another. By 1934, however, both the Arabs and the Mandatory Government had become convinced that there was no prospect of the conditions of the concession being fulfilled, and it was transferred to a Jewish settlement company. Planning, however, required a number of years, and then the Second World War intervened, holding up all further progress. It was, in fact, only after the establishment of the State of Israel that the first stage, involving deepening the bed of the Jordan for a distance of rather less than three miles south of Lake Hulah, was taken in hand. Stage B,

[1] M. Paz, *Anti-Malaria Work*. Five Years of the Upper Galilee Regional Council, 1955, pp. 30–4.

[2] Source: Drainage and Development of the Hulah; collected and edited (on the basis of various publications) by the Jewish National Fund and the Hulah Authority, 1960. 19 pp.

after four years of planning, was executed by the American Construction Aggregates Corporation.

The work was completed in 1959. Lake Hulah had disappeared from the map. The region has already benefited and stands to benefit still more from this great project which has changed the face of the landscape. But there remain a number of problems to which adequate solutions must be found. (To avoid any misunderstanding it must be stressed that the project was planned and carried out by national settlement bodies without any connection with the regional council. The settlements of the district, nevertheless, have a vital interest in its progress and success.)

Owing to defects in planning and execution which could not be foreseen, certain alterations and supplementary features must be introduced in the original plan, for the river is very much alive and a lengthy period must pass before conditions are stabilized and equilibrium is achieved. Maintenance, seasonal cleaning and repair of the channels and installations both before and after the winter flood waters will always be necessary. There will always be breaches to be repaired, blocked canals to be cleared, erosion, weeds, rodents, fires, peat deposits to cope with. The Government and the settlement institutions responsible for planning and execution undertake the repairs that come within their sphere. But maintenance will always be the function of the region.

SOIL CONSERVATION

The object of soil conservation is preservation of the soil's fertility as the basis of long-term agriculture, in contrast to exploitative farming and nomadism. It is achieved by such means as prevention of erosion and floods; changes in methods of cultivation and irrigation; soil improvement (drainage, sown pastures, afforestation and the prevention of grass fires).

Government and public institutions undertake these functions, but in this present case the main burden falls upon the settlements, and the regional council which must find the necessary funds and integrate these tasks into its efforts to provide work for the unemployed.

WATER RIGHTS SETTLEMENT

The waters of the Hulah Valley are today concentrated in three projects: the Hasbani-Dan, the Banias, and a number of smaller waterworks of the Mekorot Water Corporation. Each has its own specific problems to grapple with. The volume of water in the Hasbani brook is subject to seasonal variations, reaching its lowest point towards the end of summer. It has been found necessary to link it up with tributaries of the Dan brook, which has a more stable flow. Hence the name—Hasbani-Dan project. Fourteen settlements are associated with this project, which supplies them with many millions of cubic metres of water for irrigation and other purposes.

Five settlements are associated with the other project, while the function of the third is to supply water to the waterless southern district and to some of the highland settlements. New problems have emerged after the draining of the marshes. Power stations are being erected. All these questions call for centralized and specialized planning, and suitable local arrangements. An important rôle is reserved for the regional council in the settlement of all these problems.

Drainage

Intensive irrigated cultivation requires that the water table be kept as low as possible, no less than two metres below the surface and for certain crops as much as three metres. A higher water table prevents proper ventilation for the crop, causes salination of the soil and other difficulties.

The primitive methods of irrigation practised over the ages, the existence of the marshes and the lake aggravated local conditions. When the lake was emptied conditions were created for the draining of the entire valley.

In 1954 the Hulah Valley Drainage Authority was set up, upon the initiative of the regional council and with the active assistance of the Ministry of Agriculture and the Drainage Division of the Israel Water Authority (Tahal). After four years of surveying, research and planning, sixty-six kilometres of open canals were excavated, bridges and aqueducts were constructed and many kilometres of pipes laid. Operations are continuing, the pace being set by the availability of money to finance them. Up to the present funds have not been adequate. The main burden rests upon the settlements which are actively assisted by the council.

ELECTRICAL POWER

The council initiated and, with aid from the Upper Galilee Settlements' Association, financed the final phase of the regional electrical power project enabling the linking of all settlement waterworks and factories to the grid. In the middle of 1961 the regional council concluded its task as regional contractor of the Electricity Corporation.

FARMING ESTATES

The Deganim Ba'hulah Farm

In the autumn of 1957 the council executive mooted the establishment by the settlements of the district of a joint farming estate on
the drained area of the Hulah.[1]

After a trial period of one year the northern lands were transferred
to the Hulah Valley Authority, which we shall discuss later. The
eastern sector of the former marsh was handed to the regional
council by the Jewish National Fund, to be cultivated jointly by the
settlements. This sector was 7,500 dunams in extent. A limited company Deganim Ba'hulah was floated by the council and twelve settlements to undertake amelioration and development work.

All the settlements hold equal shares in the company, their obligations and privileges being set forth in the articles of association. The
main obligation is to supply farm machinery such as tractors and
harvesters. Only the settlements benefit from the crop which is divided
equally among them. The company keeps accounts of work invested
by each settlement while the latter pays for grain received at prices
fixed by the executive.

The settlements stand security for each other.

The national settlement authorities retain the right to settle new
settlers on the land whenever they wish to do so, provided they repay
the investments made by the company.

The name of the company indicates its main functions (Hebrew:
deganim = grains). It grows only grain and not multiennial crops
such as fruit.

The Fruit Farm

This farm was also established—and is owned—by the regional
council in 1957, upon the suggestion of Experiments Committee, to
serve as a model farm, and to conduct horticultural experiments for
the district.[2] The land here is rich and fertile and is bountifully
watered. In the 1950's apple-growing developed as the main branch
of farming. Local growers are convinced that their success is largely
the result of the high standards achieved by the regional nursery and
the joint fruit farm.

The farm is divided into two sections: the orchards which are
run on a normal commercial basis, and model and experimental
holdings. The planted areas are as follows:

(1) Apples: 90 dunams of commercial orchard and 68 dunams for

[1] Broshi, 'The Deganim Ba'hulah Farm, Upper Galilee', Regional
Council Bulletin, No. 6, August, 1960, *et al.*
[2] Simha Ziv, Bulletin of the Regional Council, No. 7.

experimental purposes. The purpose of the latter is to test distances between trees, the grafting of different varieties on different stocks, methods of cultivation, use of fertilizers, etc.

(2) Pears: 14 dunams, of which only a small part are for marketing, to serve as a model while the rest is for experimental purposes, to test varieties and stocks.

(3) Vines: 52 dunams for commercial vineyard and 23 dunams for model purposes and to experiment with methods of cultivation, selection of suitable varieties, materials for trellising—iron-piping, concrete or wooden poles—and insecticides, etc.

(4) Bamboo: a nursery to meet the settlements' demand for supports in the orchards and to develop various local industries.

(5) Carobs: to find a variety which will not shrink too much during ripening and meet local needs for a livestock feed.

(6) Pecan nuts—selection of suitable varieties.

New sections are developed every year, and recent additions include a collection of citrus varieties on various stocks, of avocados, etc.

In 1961 the farm covered an area of 354 dunams (88½ acres). The investment (up to 1960) totalled IL. 215,000.

In the planning stage is a 100-dunam citrus grove for both commercial and experimental purposes.

The Regional Nursery

The regional nursery was established in 1952 by growers in the district and the regional council. Its function is to supply vine stocks and fruit seedlings to the settlements in the neighbourhood, the surplus being marketed through the normal channels of the Tnuva Co-operative. The nursery covers an area of 400 dunams (100 acres) and includes a mother vineyard and orchard of deciduous varieties as well as a plot for growing seedlings which is rotated yearly with green manure and unirrigated crops.

The nursery has a twofold function: raising of good quality seedlings and ensuring local material for reproductive purposes.

The Experiment Station

The youngest project sponsored by the regional council is the agricultural experiment station, which was founded in 1954. The station has known many vicissitudes in its comparatively short career, and in 1959 the possibility of closing it down was considered as for a number of years it had produced no results. Considerable deficits had been incurred, the original location had been found unsuitable and changed, while the directors had been replaced several times. In spite of all these setbacks, however, it was resolved to maintain the station.[1] By 1961 the area it occupied had grown to 1,200 dunams (300 acres). The directors and personnel had surmounted the difficulties encountered, and in 1960 the highest yields of cotton, ground-nuts and potatoes in the district were obtained from it. The station accordingly had a considerable income and was able to continue with its experiments.

Tests conducted were mainly in the cultivation of cotton, autumn potatoes, ground-nuts, lucerne and maize. The station also experimented with the optimum volume of water required for various crops and irrigation seasons in connection with various agro-technical problems of the amount of water in the soil. Tests were also made to establish the equal distribution of the water quota over the area by optimal pressure in the vents along the entire length of the sprinkler, by proper spacing between the lines and the sprinklers, by conducting irrigation when there is no wind, etc. Other directions in which experiments are being conducted are in the reduction of the water ration without affecting yields and adjustment of the water régime to ensure simultaneous ripening and a single picking of the cotton crop.

The Cattle Ranch

The areas of stubble and of other offal suitable for feeding cattle which remain after the grading of grain raised the question of the profitability of breeding beef cattle. Feed available in the *summer* is sufficient for the maintenance of 3,000 head. The problem of feeding the cattle in the winter months would be solved by planting 17,000 dunams (4,250 acres) of pasture land in the neighbourhood.

The council entered into partnership with a group of South African investors, and in 1958 operations commenced with the purchase of 200 head of cattle of the local breed. At the beginning of 1961, 600 heifers of the Syrian breed were brought from Turkey in two consignments, while another 300 calves for fattening were purchased in Arab villages in Israel, bringing the herd up to 1,100 head. Two hundred heifers were artificially inseminated from Hereford bulls; the imported cattle have become fully acclimatized and the ranch seems well set on the road to success.

At the beginning of 1960 the council's four farming estates already showed a net profit.[2]

[1] Report of the Field Crops Experiment Station. Bulletin of the Regional Council, No. 11, *et al.*

[2] Bulletin of the Regional Council, No. 16, April 1961, pp. 5–6.

PEAT

Research into the possibilities of the agricultural exploitation of the peat deposits in the Hulah Valley began soon after the concession was acquired. The investigations did not produce the results looked forward to. Nevertheless financial and other aid extended by the regional council helped to find a solution to the technical problem of pumping the raw peat, and it began to be utilized in Upper Galilee and the Jordan Valley. For many crops the use of peat, fortified by chemical fertilizers, gave the same results as stable manure.

OTHER FUNCTIONS

The regional council has an amphitheatre in which theatrical and operatic performances, concerts, etc., are held. In addition there is a regional choir set up upon the initiative of the council executive. Other activities include management of the local youth hostel set up in conjunction with the Youth Hostels Association, and co-operation with the security authorities in maintaining the Frontier Force.

AGRICULTURAL SERVICES

In addition to its normal local government services the council has an agricultural department which collaborates closely with the Ministry of Agriculture, the Experiment Station, the Regional Development Officer and various committees of specialists in a number of fields such as the leasing of lands, water and field services, drainage, plant protection, machinery and equipment, labour efficiency research, the guidance bureau and laboratories. The council's crowning achievement in this sphere, however, is the farming estates which we shall describe briefly.

Estates	Expenditure (IL.)	Income (IL.)	Loss (IL.)	Profit (IL.)
(1) Fruit Tree Nursery (sales of seedlings)	116,820	113,484	3,336	
(2) Orchard Estate (sales of fruit and vine stocks)	11,213	11,788	—	575
(3) Field Crops Estate	302,649	320,081	—	17,432
(4) Cattle Ranch	23,929	21,319	2,610	
	454,611	466,672	5,946	18,007

Up to 1961 a sum of IL. 192,000 had been invested in the field crops estate.

HULAH VALLEY AUTHORITY

The completion of the first phase of the draining of the Hulah marshes emptied the latter of the standing water, dried out the areas of peat and some of the cultivable lands.[1]

The area of the marsh and the lake, 40,000 dunams (10,000 acres) in extent, which had never been cultivated before was allocated as follows:

(1) Additional land for existing settlements	5,000	dunams
(2) Wild life reserve	3,000	,,
(3) In reserve for three new villages (in the interim cultivated by the Deganim Ba'hulah Co.)	9,000	,,
(4) For cultivation by Hulah Development Authority	23,000	,,

In spite of the fact that in Israel managerial farming estates are regarded with a jaundiced eye, the following considerations operated in their favour in the Hulah.

(*a*) The danger of the rapid spread of fire, and the difficulty of extinguishing such a conflagration in the dried-out peat area.

(*b*) Special equipment, underground irrigation and large tracts of land are necessary in preparing the peat areas for cultivation. Other requisites are sound organization and rapid execution of the various tasks.

For these and other reasons the decision was taken in 1958 to set up the Hulah Development Authority as a limited stock company in which the Israel Government, the Jewish Agency for Israel and the Jewish National Fund held shares in a ratio of 5 : 3 : 2, respectively. The Board of Directors was set up accordingly.

The purposes of the company are: to cultivate crops replacing imports or suitable for export; specialization in large-scale branches; mechanization; to foster agricultural industry for the primary local processing of produce; very intensive cultivation to provide work for large numbers of immigrants.

As the event proved, the company had not been mistaken. Immediately after operations were begun fire broke out and covered tens of thousands of dunams of peat which had dried. The blaze continued for five months until, by blocking the southern outlet of

[1] Source: Draining and Development of the Hulah; a collection of articles, etc., issued by the Jewish National Fund and the Hulah Authority 1960, p. 10, *et al.*

the lake and of the inlet of Jordan to the north of the marsh, assisted by the winter rains, it was extinguished.

Twenty-three thousand dunams of the total area of 40,000 are planned to remain in the hands of the Authority for a period of fifteen years.

Agricultural and industrial plans are based upon manpower available in the area.

Six industrial installations have already been established: (a) a mint oil extraction plant by the Frutarom Co.; (b) a ground-nut sorting and grading plant by the Hulah Authority; (c) a cotton gin—in partnership with the regional council and the Ground-nuts and Cotton Marketing Company; (d) a dehydrated vegetables factory (in the town of Hatzor) in partnership with the Tiuss Corporation (a joint venture of the Israel Government and the General Federation of Labour); (e) a plant for cleaning, drying and loading various grains; and (f) a factory for grading, packing and cooling asparagus —by the Hulah Authority.

In the course of six years of operation many achievements have been registered. Not only the landscape but the very aspect of civilization has been changed. With the exception of the Hellenistic era, and perhaps also that of the Crusaders, this was a backward and sparsely settled area. Only ninety years ago there were no more than nine villages with 650 famished and fever-stricken inhabitants. Today—and the Valley is still far from being fully developed—it has eighteen settlements with a total population of 25,000. The drainage project has literally altered the physical map. The marshes and the lake have disappeared. Instead of the long, winding brooks and streams there are now straight canals, thriving settlements, verdant irrigated fields, roads and bridges, all of which are novelties. A large sum has been invested in the Valley but it has already brought forth fruit.

No one who did not see the area before the work of drainage was taken in hand can possibly believe that here was once a lake and that the sluggish Jordan fed the lush and dense undergrowth of the marsh. The wild life reserve remains, a lone witness to the past. All the marsh areas have been ameliorated and have been brought under the plough. The yields registered exceed national averages.

In 1960 crops grown on the former marshlands included cotton, ground-nuts, thyme, flowers, asparagus, vegetables, rice, maize, sorghum, winter cereals and elm trees. The cropped area was 24,750 dunams in extent; by 1964 it is planned to be 27,000 on a physical area of 23,000 dunams.

Proper cultivation will obviate any future menace of fire.

95

GALILEE DEVELOPMENT COMPANY

The economic projects surveyed up to the present were launched by diverse bodies for equally diverse reasons, with virtually no connection between the enterprises.[1] In the council's fifth year it inductively realized the need for a comprehensive economic corporation, a sort of holding company, to plan and to work for more system in the development of the district. The excellent prospects for both agriculture and industry were appreciated.

The Galilee Development Company set itself the following objectives: (a) the establishment of producers' agricultural service enterprises, including an apple packing house, a cold storage plant, a lucerne meal mill, a fruit-tree nursery, an agricultural garage, etc.; (b) consumers' services such as a bakery, etc.; (c) industrial plants utilizing locally available raw materials—a fertilizer factory to exploit the peat fields, a gravel mill, etc.; (d) plants to process farm produce—a cotton gin, lucerne dehydrating plant, fruit and vegetable canneries, slaughterhouse and freezing plant, etc.

The Company has not formulated any other aims, but there are opportunities for the development of other industries not necessarily dependent upon local raw materials the products of which can be marketed elsewhere in the country and even exported.

The Development Company was floated as an independent corporation with a basic share capital of IL. 100,000, subscribed as follows: the regional council—IL. 25,000 preference shares; Upper Galilee Settlements (the purchasing organization of the settlements in the area)—IL. 25,000 preference shares; and the agricultural villages—IL. 50,000 ordinary shares. The Company is managed by representatives of the regional council and Upper Galilee Settlements,

Every enterprise in the region constitutes a separate economic and legal body, with the Galilee Development Company serving as a holding company. The Company has laid down a number of basic principles:

(1) The benefits to be derived by any settlement from a subsidiary of the Company shall be directly in keeping with its investment and participation in management.

(2) The settlement's membership in any subsidiary is conditional upon membership in the Company.

(3) The Development Company shall have the deciding vote in every subsidiary, in order to safeguard its public character.

[1] Sources for data on 'Galilee Development Company: Five Years—Fourteen Projects': a pamphlet of the Upper Galilee Regional Council. 99 pp. Also: 17 bulletins of the Council for the years 1957-61. Other data were derived from current reports and balance sheets of the Council.

96

(4) The Company may seek public or other investment in its enterprises.

In 1955 when the Galilee Development Company was floated the area was already being developed at a rapid pace, and the notion of establishing joint enterprises, in which the settlements of the region would hold shares, had already gained considerable support.

The entry of the settlements into the industrial field raised the question of ownership of the enterprises. After much discussion a method of co-operative ownership by the interested settlements, with the participation of the Galilee Development Corporation and other, external, bodies, was decided upon. Thus, to cite one example, the partners in the spinning mill are: six settlements, the Galilee Development Company, the Co-operative Centre of the General Federation of Labour and the Moller Textile Corporation. The structure of ownership of the local cotton gin is different: the proprietors are fifteen settlements, the Galilee Development Corporation, the Hulah Valley Authority and the Ground-nuts and Cotton Marketing Company. The partners of the Cold Storage plant are: sixteen settlements and the Galilee Development Company

This, however, is not the only model of ownership, and the region does not exclude the possibility of sole ownership by the holding company.

The industry of Upper Galilee, it may be said, has been developed without any central plan. In no small measure it has been a spontaneous growth, which has also, here and there, given certain play for individual and collective initiative. Some enterprises were established under the impact of the Government's siting policy, within the general context of national economic and settlement planning.

In spite of this, however, the siting is fortuitous. Here as in other regions the location of any project is largely dependent upon the alertness of those sponsoring and establishing it. Almost all regions suffer from the absence of preliminary planning. Many settlements occupy sites that are not suitable, and even in the location of new towns chance has been a major factor. Today, of course, it is difficult to alter established facts; this, however, is a subject to which we shall return at the end of the present chapter.

JOINT ENTERPRISES IN THE REGION

1. *Galil Ice and Cold Storage Co. Ltd.*

This concern was founded in 1954 in Kiryat Shmonah. The partners are as follows: The regional council (20 settlements)—50 per cent, and the Co-operative Centre of the Histadrut—50 per cent. The original investment was IL. 140,000 and the capacity of the plant was 250 tons. In 1960 the plant could store 2,000 tons of fruit while the capital investment had grown to IL. 1,250,000. Immediate objectives are to increase capacity to 3,000 tons and to invest a further IL. 600,000. In the planning stage is another cold storage plant to serve the southern area (to be erected in Hatzor). This plant, in which IL. 1,200,000 will be invested, will have a storage capacity of 1,000 tons of fruit.

2. *The Regional Packing House in Kiryat Shmonah* sponsored by the regional council. The building covers an area of 737 square metres, in which, together with the equipment, a total sum of IL. 200,000 had been invested by the close of 1960. The expansion of the packing house to enable it to handle 4,000 tons of fruit is being projected while the construction of another one to serve the southern section and to be erected in Hatzor is planned.

This packing house, established in 1955, has known some difficult years. The settlement could not furnish the necessary personnel in the packing season and the standard of hired workers was low. Furthermore a number of settlements preferred to undertake their own packing. The season, it should be mentioned, lasts only 40–50 days.

Over a period of two years, however, the difficulties were surmounted by installing a waxing machine which ensured that the fruit would keep better under conditions of refrigeration, the utilization of packing containers, mechanical lifting devices, etc. Loss in weight was also reduced.

The lesson learnt in the course of the early years of operation led to the introduction of the following principles when the packing house was reorganized: (a) participating settlements undertake to supply a minimum quantity of fruit for packing; (b) privileges and obligations to be calculated in keeping with the quantity of fruit supplied. This index shall be applied in investments and in the appointment of managerial personnel; (c) organizational and financial operations to be conducted independently by a board of directors and treasurer elected by the participating settlements, etc.

In 1961 the packing house operated under the direct ownership of the settlements. It packed 2,000 tons of fruit as against 150 tons in 1955. Today eleven settlements are associated in this venture, eight of whom use its facilities to pack their entire crop, while the remaining three pack only part of their crop in it. Picking is carried out with the aid of containers drawn by tractors which transport them straight to the packing house. The containers, which are upholstered to protect the fruit, are raised to the packing machine with the aid of a hydraulic lift. The fruit which is to be stored for a lengthier period is waxed, thereby preventing loss of weight. All fruit accepted leaves the packing house on the same day; most of it is placed in cold storage in an adjacent building while the balance is stored in Tel Aviv.

Third grade fruit is despatched to the canneries.

The plant employs about one hundred workmen in all.

3. *Pri Hagalil Ltd.* in Hatzor for canning and dehydrating farm produce. Three bodies collaborated in the flotation of this company in 1960—The Tiuss Corporation, the Upper Galilee Regional Council and the Hulah Valley Authority. Fifty per cent of the shares are owned by the first two and the remaining 50 per cent by the Hulah Valley Authority. Fourteen settlements participate. The initial investment capital was IL.1·1 million. The board of directors appoints the management from among its own members. The production programme includes: (*a*) products for export; (*b*) absorption of surplus farm produce; (*c*) establishment of equilibrium in agricultural and industrial employment. The decision to establish the enterprise in Hatzor was taken in deference to the wishes of a number of members of the Government.

The first season's programme envisaged the dehydration and canning of 4,000 tons of produce, intended for export. Fluctuations in the supply of vegetables, however, did not permit this goal to be achieved. The canning department processed melons, figs, apples, plums and pickled cucumbers, while the dehydrating section produced dried onions, tomatoes, apples and figs. In the first year of operation the plant employed 50–80 workers which rose to 150–250 in the second.

One million pounds have been invested in this enterprise, 10 per cent of it coming from the settlements. Output in 1961, the second year of production, totalled IL. 2·5 million. According to initial reports from the United States, Pri Hagalil products are competing successfully not only with similar canned goods from other countries but with high grade produce from California.[1]

4. *Manpetat Hazafon Ltd.*—Northern Cotton Gins.

Cotton growing has undergone rapid expansion in the Hulah Valley in the past few years, totalling 17,000 dunams (4,250 acres) in 1960, thereby making necessary the establishment of plants to process the crop. The shareholders of the gin are as follows: the settlements of the region—33 per cent; the Hulah Valley Authority—33 per cent; and the Ground-nuts and Cotton Marketing Company—34 per cent. The investment capital of the gin is IL. 1·2 million. Nineteen

[1] Bulletin of the Upper Galilee Regional Council, No. 18, August 1961, p. 12.

settlements participate in the plant which went into operation towards the close of summer 1959. The number of workers is 27.

There are three managers representing the shareholders. The plant began with two gins capable of ginning 60 tons in 20 hours, but another two gins will be installed to meet the needs of the growing area under cotton.

5. *Matviah—Hulah Textile Ltd.*—Hulah Spinning Mills.

It proved very difficult for the regional council to interest investors in this project. Even the local settlements were reluctant. The apprehensions entertained were indeed not unfounded as a project of this kind calls not only for a very considerable investment but also for much technological know-how. The regional council indicated its willingness to take up a 25 per cent interest in this spinning mill through its Galilee Development Corporation. After protracted negotiations the Moller Textile Corporation agreed to subscribe 50 per cent of the capital, after which the Co-operative Centre of the Histadrut took up the remaining 25 per cent.

The mill is planned for 10,000 spindles, and the investment capital required is of a magnitude of IL. 3·75 million, of which the Government will provide 70 per cent, and the partners the balance of IL 1·1 million. The mill will produce uncombed yarn. In full production it will provide employment for 200 workers.

The Galilee Development Corporation is backed by six settlements each of which has undertaken to invest IL. 35,000. In addition each of them will furnish 4–5 workers (including workers of limited capacity). It is noteworthy that, notwithstanding the heterogeneous composition of the partnership, relations within the plant are harmonious and satisfactory progress has been registered.

6. *Of Hagalil, Ltd.*—Galilee Poultry.

This is a regional enterprise for the slaughter and plucking of poultry, with a refrigeration and deep freeze plant attached. For some reason co-operative marketing bodies have not entered this field.

Poultry breeding has undergone very rapid—perhaps too rapid—expansion in recent years, and prices for live poultry have been subject to sharp fluctuations. The slaughterhouses and cold storage plants have helped to regulate the supply. For the Upper Galilee, situated at a considerable distance from its markets, such a plant, which reduces costs of transport and loss of weight, is of particular importance.

This enterprise, constituting a partnership of twenty settlements, differs radically from others in the district. In the course of time, it may be expected, other settlements will join.

100

Refrigeration equipment includes a freezing room at a temperature of $-40°$ C., two cold storage rooms at $-18°$ C., and one at $0°$ C. Daily capacity is 5 tons, while deep freeze storage capacity is 50 tons. So far IL. 250,000 has been invested and another IL. 300,000 will be necessary to finance further expansion.

A cattle abattoir to serve the district is also planned; it will call for an investment of IL. 500,000.

7. *Mehzavei Hagalil Ha'elyon*—Upper Galilee Quarries.

Following the location of a suitable site for a quarry to serve the local settlements the Galilee Development Company went into partnership with a transport co-operative to exploit deposits of stone. Local production of gravel and ground sand for building purposes has effected a considerable saving for the settlements of the region.

8. *The Regional Agricultural Garage*, for the repair of tractors and other agricultural vehicles, was established by the transport co-operative and a garage. Almost all of the settlements of the district are members of the transport co-operative.

9. *Lahmenu*—Our Bread.

This regional bakery was set up in 1953 in conjunction with Hamashbir La'oleh to ensure regular supplies of bread for the town of Kiryat Shmonah and for the settlements in the vicinity, thereby relieving the latter of the need to bake their own bread. However, when after the passage of a number of years relations between Kiryat Shmonah and the regional council became aggravated, the former established its own bakery. (We shall deal later with relations between the regional and local councils.)

This comparatively large number of important enterprises were established in a short period and in close succession. Many other regions in Israel are still planning similar enterprises, and there can be no doubt about the importance of the example afforded by Upper Galilee. The latter is now entering upon a phase of horizontal and vertical expansion, to consolidate the undertakings established in so short a time.

For example, in 1961 the capacity of the cold storage plant was expanded to 1,350 tons. Pri Hagalil is erecting a fruit juices section in addition to building new stores. The cotton gin is being equipped with a delinting machine, to undertake the first stage in extraction of oil from the cotton seed. Finally the spinning mill is being doubled.

The regional council, indeed, has already proved itself, and there is no major project in the area on which it is not consulted or in which it is not directly interested. It has also initiated the introduction of new crops, the processing of local raw materials, etc. In fact a tradition of large-scale and basic regional co-operation is steadily being developed.

ENTERPRISES IN THE DISTRICT NOT OWNED BY SETTLEMENTS

There are seven factories—not workshops—in the area established by public and private enterprise in which neither the Galilee Development Corporation nor the settlements have any interest. Tnuva, the national agency for marketing farm produce, has a district creamery in the *moshav* of Beth Hillel. The Sollel Boneh–Koor concern (of the Histadrut) operates three undertakings in this neighbourhood: the Ramim (Vulcan) metal works in Kiryat Shmonah; the Even Vasid quarries (in partnership with the *kibbutz* Lehavot Habashan; and the Hemar concrete works, also in Kiryat Shmonah. There are three privately owned enterprises: 2 diamond polishing plants in Kiryat Shmonah and Hatzor, respectively, and the Hermon Bakery in Kiryat Shmonah.

INDUSTRIAL ENTERPRISES IN AND OWNED BY THE SETTLEMENTS

There are 25 industrial enterprises in this region owned by 21 of the 33 settlements. They are engaged in the following branches: metal-work—8; carpentry and joinery—2; machinery—2; agricultural and other equipment—6; footwear—1; lucerne drying—3; plastics—1; quarrying—1; brushes—1.

The plants are not of large proportions and average no more than 15 workmen. The sole exception is the Neot Mordechai shoe factory which employs 60–70. Thus these enterprises are on the borderline between workshops and factories. They can best be classified in the category of petty industry. One plant which, because of the branch of industry in which it engages and notwithstanding the small number of workers it employs (10), must be regarded as a factory is the Mehaprim works in Kfar Gileadi, producing hydraulic equipment for earth moving, building and elevating. The settlements' principal motive in establishing this plant was to create an outlet for the youth, who take gladly to metalwork. Mehaprim produces two types of excavators—ditch-diggers and pit-diggers; two types of front loading attachments for wheeled tractors; and two types of 'shoveldozer' loaders for crawling tractors. Other products include a wide variety of auxiliary equipment for all types of tractors, scrapers, cranes, etc.

For the time being, because of their dimensions, these plants do

102

not constitute any special problem for the settlements or for regional industrial planning. In the event of any enterprise proving burdensome or unprofitable for any settlement, the decision is taken to liquidate it—as has already happened in two cases. Should it, however, grow beyond the operational capacity of the settlement, a partnership may be established with others, as was the case with the first lucerne drying installation, which was undertaken by five settlements, in addition to the establishment of another by six other settlements. In 1961, it may be noted, a third was set up by a partnership of six *kibbutzim*.

In the light of the patent willingness of settlements to form partnerships to operate larger enterprises and particularly in view of the existence of many such partnerships, there does not seem to be any insuperable difficulty in the way of setting up other regional projects. Indeed in Upper Galilee where the *kibbutzim* make up the majority of the settlements, inter-*kibbutz* collaboration in various undertakings on their own account and responsibility is general. The regional council lends its assistance in organization and in negotiation.

MOSHAVIM

The eight *moshavim* in Upper Galilee—six of which are young settlements still under the tutelage of the Jewish Agency's Settlement Department—have not demonstrated a similar degree of readiness to integrate themselves in this region. These *moshavim* have 1,800 inhabitants as compared with 8,000 in the twenty-three *kibbutzim*.

The regional council has extended very substantial assistance to the *moshavim* in the installation of electricity, maintenance of schools and kindergartens, school kitchens, veterinary services, sanitary inspection, through its technical department, in the supply of bread, and in the organization of cultural and sporting activities.

As long as the younger *moshavim* remain under the supervision of the settlement body they will not evince any regional initiative. Once they achieve managerial and economic independence and are self-supporting, they may be expected to be drawn, more and more, into the life and activities of the district.

SUMMARY OF REGIONAL CO-OPERATION

In the daily affairs of the region and regional council all settlements have been co-operating for many years in a spirit of complete harmony; differences of opinion stemming from party politics have hardly ever manifested themselves. The homogeneous economic and

A. —H   103

social background has served as a powerful unifying cement, despite political heterogeneity.

Nineteen co-operative enterprises in the district either come within the scope of the regional council or the latter holds an interest in them, or are based upon various link-ups between the settlements. They are as follows:

(a) Agriculture
    (1) Deganim Estate, Hulah Valley
    (2) Joint fish stores
    (3) Nursery and fruit farms ⎱ of the
    (4) Field crops farm   ⎰ regional council.

(b) Industry
    (1) Three lucerne meal mills (partnerships)
    (2) Apple packing house
    (3) Cold storage (Kiryat Shmonah)
    (4) Poultry slaughterhouse
    (5) Spinning mill (partly owned by the settlements)
    (6) Cotton gin (partly owned by the settlements)
    (7) Drying and canning of fruit and vegetables (in Hatzor)
    (8) Transport and automobile repair co-operative
    (9) Cold storage and packing house (in Hatzor)
    (10) Bakery.

(c) Waterworks
    (1) Hatzbani-Dan
    (2) Banias
    (3) Malha (for settlements in the southern area).

(d) Services
    (1) Purchasing organization
    (2) Heavy machinery (tractors, and excavators for earth moving).

The list, it will be seen, does not include cultural activities, sanitary services, etc.

The problem engaging the district is that of vocational and post-secondary education. In the *kibbutzim* all young people are given a secondary education (this is the only youth of the working class in Israel which is) but few of them go on to a higher stage such as a teachers' college, the Israel Institute of Technology and the Faculty of Agriculture of the Hebrew University.

Normally young people in the *kibbutzim* learn their trades on the job, perhaps with some—intermittent and insufficient—theoretical training when they are older. Their schooling, accordingly, centres about national values, settlement, security, the labour and the *kibbutz* movements. They are lacking in higher education.

In the Jordan Valley there is a single vocational school of secondary standard. In Upper Galilee local leaders are considering the establishment of a post-secondary school.

TOWN AND VILLAGE IN UPPER GALILEE

With the exception of the city of Safad, which is set apart from its environment, Upper Galilee has always been an agricultural district. Thus when in 1949 the Government Planning Department proposed the development of a new town on the site of Arab Halsa, a searching discussion ensued in the *Vaad Hagush* (Bloc Committee) of the Upper Galilean settlements. Some favoured the establishment of a *moshav* instead. In the meantime, however, the Jewish Agency repaired a number of houses abandoned by the Arabs, in which it housed a few dozen immigrant families coming from Yemen. The latter were to engage in primary development work, and in the manufacture of reed mats, under the supervision of certain of the *kibbutz* settlers.

The Gush Committee was reluctant to take any decision on the future of Halsa and so conducted a referendum in all local settlements. The latter called general meetings at which the matter was carefully deliberated. Finally the decision was taken to agree to the urban settlement as proposed. In retrospect, it seems that the decision may have been mistaken, as the district was hardly ready to establish an integral town in its midst. It might have been better had the newcomers been settled in two or three—or even more—*moshavim*. No one, however, can be blamed for taking this fateful decision. There is no question but that the motives were of the best. What was lacking, however, was long-sighted planning.

The new town began to absorb new settlers at a rapid pace until they constituted a majority of the inhabitants of the district.

When the regional council was inaugurated in 1950 it set up its offices in Halsa. The council's immediate and main concern was to provide employment for the immigrants on local development works. They were also employed in the agricultural settlements, on local building construction, on the Hulah drainage project, afforestation, in the garage, the cold storage plant, the bakery, the workshops and the various institutions. The council's functions included the supply of water and street lighting, roads, social welfare, health, education, security, religious services and parks. A Kiryat Shmonah (as the town was renamed) development company to sponsor new enterprises to provide employment for the new inhabitants was set up.

Government funds and loans to finance the absorption of the newcomers and the provision of employment for the workless

105

facilitated the expansion of the agricultural area and the improvement of local services. New industries were established, their main
purpose being to provide work for the citizens of Kiryat Shmonah.
At this stage the settlements were animated only by national considerations in the aid they extended to the young town. Later, they
saw the prospects of establishing enterprises of which the settlements
stood in need for economic reasons.

The villages had been given a new stimulus to intensify, still
further, existing branches of agriculture, such as various industrial
crops, the planting of new orchards, and development of a nursery
and experiment station. Few of these projects would have been
undertaken had it not been for the need to find employment for the
inhabitants of Kiryat Shmonah. In the meantime the latter, who had
been devoid of any experience of farming or even manual labour,
learnt agricultural occupations. Some of them joined new immigrant
settlements which were founded in the vicinity.

The new crops required storage and refrigeration facilities, or
must be processed before they could be marketed. The only logical
thing to do was to establish the plants required in Kiryat Shmonah.

Expansion and intensification of agriculture, even on the dimensions undertaken in the district, were not sufficient to provide a basis
for the new artificial township. Nor did the expansion of the small-
and medium-sized factories in the settlements, to create employment,
solve the problem. Inexorably the need to provide work led to the
industrial development of the district.

Private enterprise is not interested in investing in the new townships, though here private capital did make some contribution.

Kiryat Shmonah grew apace. Two banks opened branches in the
town, there was a post-office and a clinic. The 15,000 inhabitants are
an important market for local agricultural produce. It has become a
town, though the main sources of employment are still soil improvement and conservation, drainage, irrigation and afforestation. Many
inhabitants are engaged on development works, such as building and
road construction, electricity and telephones. Forty per cent of the
earners are still employed on a temporary basis. The gradual transition from development works to more permanent employment is
being undertaken by the regional council in conjunction with the
Ministry of Labour.

The regional labour force has not yet reached a state of equilibrium.
The know-how and the skilled artisans are in the *kibbutzim*; in
Kiryat Shmonah are the unskilled workmen. Many years must pass
before equilibrium between the advanced villages and the raw town
is achieved.

Over two-thirds of Kiryat Shmonah's population are immigrants from North Africa, Iraq and Persia; less than one-quarter come from Eastern Europe and less than one-tenth are established residents.

It is difficult to conceive the future course of Kiryat Shmonah without a lengthy period under the guidance of the settlement bodies and the Government and the constant influence exercised by environment. But in the meantime bitterness has been engendered and a dangerous antagonism has developed between the town on the one hand and the environment and the regional council on the other. This is the sort of situation which demagogues can only too easily exploit.

For its first four years, Kiryat Shmonah was represented on the regional council. In 1953 it was granted the status of a local council, and recognized as a municipal enclave within the jurisdictional area of the regional council of Upper Galilee.

The Ministry of Interior granted similar municipal status to five other settlements in this district who demanded it: Metula, Yessud Hama'alah, Hatzor, Rosh Pina, and Amiad. Thus there are six municipal islands, or rather enclaves, within the area of the regional council.

One of the most difficult tasks within human experience is to alter maps. But perhaps Time will do what administrative considerations did not succeed in doing.

The town of Hatzor still lacks a foundation. Indeed its very establishment was an error. The site of an ancient town and a group of dilapidated buildings for some reason caught the eye of the planners, and without sufficient preliminary study a new town was set up.

For the time being Hatzor's very existence is problematical, but if nevertheless artificial respiration should infuse life and vitality into it, it will have to cope with the same difficulties confronting its older sister. Logically effort and available managerial and other manpower —which in any case is limited—should have been concentrated upon one town, possessing reasonable prospects for the future, rather than divide them between two—in both of which so much is still lacking. The second town in the district possesses no functional advantage, not even from the point of view of communications and transport. Quite simply, it is redundant and as such is retarding the progress of its neighbour.

Hatzor and particularly Kiryat Shmonah jealously maintain their separate and independent existence. The link with the regional council has therefore become attenuated, but in the meantime new contexts for collaboration such as the regional town planning commission, the veterinary service and the passenger station near the local flying field have been created. It is relevant to mention here that

107

the regional council in conjunction with the Safad town council was instrumental in introducing air services. The dynamic process of development daily raises new problems which can only be solved by co-operation.

The authorities, that is to say the central and district government, do not administer or develop any new towns. For this purpose some economic basis is necessary. In the case of Kiryat Shmonah the agricultural hinterland and the local fund of experience, know-how and enterprise provided such a basis; the town should have continued as an integral unit of the environment which gave it birth after so much pain and difficulty. It is in the nature of a parent to train his son to fend for himself. In one hundred, fifty, thirty years or even less Kiryat Shmonah might find itself independent of its agricultural hinterland. It may develop industries unrelated to its environment as has been the case with thousands of cities throughout the world. At the present moment, however, there is still a prospect of converting Upper Galilee into an *agrindus*. At least let the planning of Upper Galilee and its new towns serve as an object lesson for those who adopt the *agrindus* idea.

We venture to think that had Kiryat Shmonah from the very outset been planned as a co-operative town, based upon the same principles as all the economic, social and cultural undertakings in the region, it would have been to the benefit of the entire district and, of course, of the citizens of that town. The reorganization upon co-operative lines of the first generation of Kiryat Shmonah and other development towns in Israel does not seem to be within the bounds of possibility. Their present citizens come from backward countries, ignorant of the very concept of co-operation or even partnership. From now on, however, we must educate the second generation in these new townships, already existing or to be established in the future. Overnight *agrindus* will not develop thews and sinews. It will have a long road to go, which must yet be charted and paved, and from which pitfalls and stumbling blocks must be removed.

TOWN PLANNING

As far back as 1955, that is when the regional council had already completed its fifth year, the council's engineer complained about the absence of any local town and village planning commission, because of the separate existence of the local councils of Metula, Yessud Hama'ala and Kiryat Shmonah, and because the regional council's area of jurisdiction was still not clear. The Hulah concession area as well as the lands of a number of settlements were still beyond its borders.

The engineer was of the opinion that the local commission should engage in the physical planning of the entire district provided a contiguous area came within its scope. He recommended the creation of a single commission upon which all municipal bodies in the area should be represented. 'While various, and sometimes opposed, internal interests do exist, co-operation will assist in finding a satisfactory solution. Should objective conditions require a review of the situation at some future date, there will be time enough for separation,' [1] he argued with patent justification.

At the beginning of 1958 a town planning commission, in which both the regional council and the local councils were represented, was set up in Upper Galilee. Other members of the commission were the District Officer, representatives of the Ministries of Health and Labour and the Government Planning Department.

An area of about half a million dunams (125,000 acres) came within the jurisdiction of this commission, which is one of the few bodies of its kind whose area includes both towns and villages.

In the rural districts—unlike the towns—there is a tradition of very long standing of complete freedom in planning and construction. Even the *kibbutzim* were not prepared to submit to this 'decree' requiring building permits. Here the position differs even from that in the *moshavim* where there is the problem of farmyard boundaries. In the *kibbutz* all property is communally owned and construction, in the majority of cases, is centrally planned and often centrally executed by the various *kibbutz* movements. However, this opposition to town and village planning cannot stand up to criticism. It is not a formal question. In the *kibbutzim*, under normal circumstances, a restricted group takes the decisions on building, and factors are created which are opposed to all objective architectural and planning considerations.

Planning in the region, for a variety of reasons, is not satisfactory; the legal appointment of the commission has not resulted in sufficient improvement in this respect. This is not only because the settlements are not accustomed to applying for building licences but because every national body still retains a free hand to operate in the region without having recourse to the town planning commission. The Jewish Agency's Settlement Department plans and establishes villages as it sees fit. The technical departments of the collective movements put up or alter buildings quite freely. The Jewish National Fund, in setting aside areas for afforestation, for nature reserves, for national parks, etc., is conscious of no restrictions. The Public Works Department plans roads, the Israel Water Authority (Tahal) lays down water pipes and drainage canals, the Shikun Housing Corporation puts up houses—each in accordance with its

[1] Five Years of the Upper Galilee Regional Council, June 1955.

own separate considerations. The Planning Department of the Ministry can—but does not—exercise its authority to approve and co-ordinate all these activities.

Government departments, under law, need not have recourse to local planning bodies. They co-ordinate their projects with the latter if they choose to do so.

The harm done as a result is greater and more apparent in the development towns than in the *kibbutzim*.

Many problems have not been solved, but in the meantime factors have been created which it will be difficult if not entirely impossible to change.

As a result of lack of foresight the industrial zone of Kiryat Shmonah is already too small and not suited to the needs of future development. It has been suggested that a new zone be laid in the middle of the Valley in the Hulah Authority's area.

In the three years of the operation of the town and village planning commission (up to 31 March 1961) plans for buildings covering an area of 119,000 square metres were approved—100,000 square metres by the Upper Galilee Regional Council and the balance by the local councils of Kiryat Shmonah, Metula and Yessud Hama'ala.

Economic and social ties in the region are still strong and effective planning is still within the bounds of possibility. The mistakes of the past may yet be corrected.

FUTURE PLANS PROSPECTS

Activities undertaken in Upper Galilee since the establishment of the State have been largely of a spontaneous character without any attempt at more comprehensive planning, but there is still room for the application of remedies and improvements, and certainly to ensure that future development of the region proceeds systematically towards the desired objective.

The favourable aspects and factors can be summed up as follows:

(1) There are thirty-eight settlements in the region; thirty-three of them are affiliated to the regional council. This is the optimal number for the project propounded in the present essay.

(2) The area of the region—half a million dunams—can serve as a basis for development even from the long-term technological point of view. The fertile soil, more than adequate rainfall and bountiful other sources of water bespeak reasonable economic stability in the future.

(3) The habit of collaboration has been developed, and considerable experience has been gained in partnerships between settlements and in the management of local and joint industrial enterprises.

110

Disadvantages and possible remedies:

(1) The enclaves formed by the local councils upset the contiguity of the area. These councils must be abolished; they can be included in the regional councils and drawn into the economic and other partnerships.

(2) Economically the town of Hatzor has no *raison d'être*. Some more effective use for the area must be found.

(3) The municipal, social and economic structure of Kiryat Shmonah has been defective from the outset. In spite of the many major difficulties that must be surmounted remedies can be found and applied, even though over a lengthier term. In the earlier phases the regional and the local councils must enter into joint ventures, in which the former must demonstrate breadth of vision and a spirit of liberality. As many as possible of the citizens of Kiryat Shmonah who are permanently employed by regional enterprises must be organized in groups (perhaps on the lines of artels) which shall be recognized as partners in the enterprises, or given representation on production councils. The co-operative enterprises in the town must be included in this arrangement.

(4) The *moshavim* of the district do not yet hold any interest in the enterprises. It is essential that they be brought in and their surplus farm labour be absorbed in these undertakings.

(5) Partnerships in many of the enterprises are of a partial character. A rule should be instituted that every settlement must be a member in the Galilee Development Corporation and also of every enterprise in the region.

(6) The proposal to develop another industrial zone in the region should be rejected.

(7) The town plan of Kiryat Shmonah does not take into account the potentialities of future development. An area encircling the town must be transferred from the regional council to provide for future urban development. Present strained relations in the area should be overcome and improved.

(8) The opinion still widely held in Upper Galilee that individual settlement should continue to operate industrial enterprises in the future is not correct. Even if we stop short of proposing the removal of existing local factories, there is no question that the view—shared by the Ministry of Interior—that new undertakings should be located in the urban centre and organized upon a basis of regional co-operation, is fully justified. This, however, must not be regarded in the light of external compulsion. It is in the interest of all concerned.

Map 5 shows existing regional boundaries and the enclaves; the changes we propose by the abolition of the enclaves are marked by a broken line.

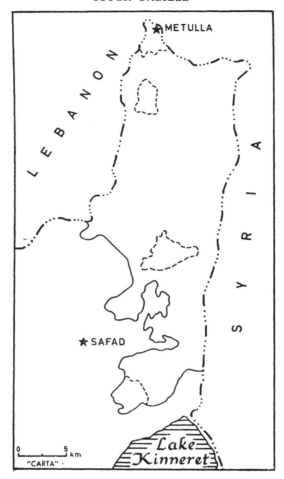

Map 5. Upper Galilee

The present amorphous character of the district is the result of the
fact that despite the achievements of those engaged in the actual
work of development, and the sincerity and goodwill of the planners,
they had before them no more definite and comprehensive objective.

With the idea of *agrindus* to guide them there is little doubt that
some method of co-ordinating the activities of internal and external
factors can be devised. The consent of the latter can be obtained
without difficulty. The future of *agrindus* depends on those within,
who must build it up and live within it.

X

THE JORDAN VALLEY

THE Jordan Valley stretches from the point where the River Jordan issues from Lake Hulah in the north along the entire course of the river down to the Dead Sea in the south. This elongated valley, together with the Arraba, constitutes part of the great rift continuing via the Gulf of Akaba down to East Africa.

The Israel–Jordan frontier runs through the northern section of the Jordan Valley (part of which is in Israel territory) and the Arraba, south of the Dead Sea. The Valley is a great depression, comprising a number of distinct sections, such as the Genossar Lowland in the north, the Beit She'an Plain in the middle and the Jordan Plain towards the south. The Jordan Valley includes only the northern section of the Kinneret (the Sea of Galilee). The Kinnaret Lowland lies about 200 metres below sea-level and has a typically sub-tropical climate. For this reason only irrigated farming is engaged in.

Soils in this district are mainly of the deep calcareous type and are highly permeable.

This combination of high temperatures and cheap water, as well as the fact that calcareous soils provide an excellent bed for irrigated crops, has assisted in the development of a very intensive agriculture. The chlorine content of the water of the Jordan is higher here than in the Hulah Valley, before the river enters the Kinneret, fluctuating between 120 and 300 milligrams to the litre. However forty years of irrigated farming in the area have not resulted in any deterioration of the soil. Bananas and lucerne are the principal local crops. Fish-breeding ponds are also favoured, more so indeed than fishing in Lake Kinneret. Before these three branches of agriculture achieved their present pre-eminence, the pride of the valley was the tomato. Today, too, the Jordan Valley produces highly profitable early tomatoes and cucumbers. Lucerne, together with a large variety of other irrigated fodder crops, has stimulated the rapid development of dairy-farming. The lack of natural grazing on the other hand has retarded sheep-raising and cattle-breeding.

In addition to the banana and table grapes, the cultivation of sub-tropical fruits, especially dates, has been introduced.

In planning the future development of the area priority will have to be given to crops whose water requirements are more modest, for local farming is already menaced by a rise in the level of subterranean water and salination of the soil.

An extensive drainage system and suitable rotation of crops to balance the use of water are essential to ensure the future of agriculture in this bountiful region.[1]

In the Jordan Valley there are twenty-one villages with a total population of 9,000. Eighteen of the settlements belong to the regional council. The three which did not join the regional council and have been constituted as local councils are: Migdal, Menahemia and Kinneret.

Here a Va'ad Hagush (Bloc Committee) was set up as far back as 1934. The Committee undertook certain municipal functions, but in the main it operated as the Jordan Valley Settlements Organization in matters of supply. When the regional council was set up in 1949 (with a jurisdictional area of 45,000 dunams of cultivated land), the Organization confined itself exclusively to supply, though on an extensive scale. The Organization operated on behalf of all eighteen settlements in the purchase of feed, seed, fuel, chemical fertilizers, building materials, clothing, footwear and provisions. In the course of a decade its turnover grew twelvefold, its own capital fourteenfold and operational capital tenfold.

The settlements of the Valley also joined hands in the establishment of a transport co-operative. The first lorry to be operated in the Jordan Valley was purchased by Kibbutz Afikim in 1929, to transport members of the *kibbutz* employed at the Naharaim Power Station. The settlements made use of farm wagons, the railroad and two privately owned lorries from Tiberias for the marketing of their produce. Gradually, however, the settlements acquired motor vehicles to despatch growing quantities of farm produce to all parts of Palestine, and even to neighbouring countries, particularly Syria. In 1938 six settlements with a combined fleet of fourteen trucks set up a transport organization. In the course of ten years the fleet has grown to eighty vehicles of considerably heavier tonnage. A garage was established, and in the course of time the need for reorganization upon a co-operative basis was recognized. The garage was moved to a new site to serve the settlements of the region. A filling station and an

[1] H. Halperin, *Changing Patterns in Israel Agriculture*, Routledge & Kegan Paul, London, 1957, pp. 11–12.

accessories and spare parts store were also established. Near by the Licensing Authority set up a testing compartment for motor vehicles and tractors.

When the question of the reorganization of transport in the region was under discussion it was generally felt that some more definitive and obligatory framework was necessary. The settlements were fully familiar with the deficiencies of transport co-operatives but, they resolved, it was the best type of organization for an enterprise of this kind.

Service enterprises in the region in 1961 were as follows: a garage; a spare parts store; heavy equipment; a banana curing house; a banana packing house for exports; a vegetable and fruit packing house also for exports; and a quarry. These enterprises employ 51 members and 155 labourers.

Industrial enterprise will be discussed later in this chapter.

THE GARAGE

The transfer of the garage to its new premises brought to its close a lengthy initial phase. Today the region is interested in developing it as a permanent enterprise by increasing the personnel, especially of highly skilled mechanics, etc., ensuring continuity of operation and the development of a department for the repair of tractors. In 1961 its staff included 10 employees and 20 members.

HEAVY EQUIPMENT

Even a large *kibbutz* is not able to maintain heavy equipment for its own use. The combination of a number of settlements for this purpose enables them to acquire and operate heavy equipment for deep ploughing, for contours for the banana groves, and for maize and asparagus, for the removal of manure from pens and the excavation and repair of fish-breeding pools.

In 1955 a regional station called the Jordan Valley Region Heavy Equipment was established for this purpose. By 1961 the inventory of the station included tractors, shovel-dozers, motor graders and large trailers. Its main function is to carry out seasonal agricultural tasks, but it also engages in various development and maintenance works in the region. In addition, in slack seasons it is employed by national construction and irrigation companies such as Sollel Boneh, Mekorot, etc.

Needless to say, continuous operation of this type of equipment reduces costs considerably. The revenue of the station, which employs fourteen workers and two members, was IL. 371,000 last year.

SPARE PARTS STORE

In 1951 the trade in and control of various commodities, especially those which had to be imported and paid for in foreign currency, was almost chaotic. At this time it proved possible to purchase surplus equipment of the American Army in Europe at low prices. The opportunity provided the stimulus for the formation of a partnership for the purchase of this equipment and especially of spare parts. At first the store was attached to the transport organization. In 1960 it was transferred to new ownership—Miphalei Emek Hayarden (Jordan Valley Enterprises)—and to new premises.

In addition to spares for motor vehicles the store also supplies various accessories, tyres, electrodes, nuts and bolts, oxygen and various tools. A department for spare parts for tractors and agricultural machinery is at present under construction. In 1961 it had a turnover of IL. 1·5 million.

The partnership has enabled the settlements to purchase at lower prices besides securing more favourable terms of credit.

The operations of the store and the garage are closely integrated in this region.

The success achieved in this field has led to the development of a number of other enterprises.

BANANA CURING HOUSE

This enterprise matures and packs 3,000 tons of bananas for the local market in the season. Experiments are constantly being conducted to find better methods of handling bananas, and especially of packing. Manual handling has been discarded and a moving belt system introduced. Piece-work on a contract basis was found to be detrimental to sound operation and a daily wage system is now the rule. At the same time there is an unremitting effort to raise daily output, to lower costs in terms of tons handled, and to find a suitable use for the offal.

BANANA EXPORT PACKING HOUSE

The Jordan Valley was the first district in this country to develop banana growing, and from the very outset sought export outlets independently. Prior to the Second World War certain progress had been registered. Israeli bananas were being sold even in Soviet Russia. After the war banana growing was reorganized, and efforts to find suitable markets abroad in Western Europe and also in Russia were renewed in 1947.

116

Since then banana growing has spread to other parts of Israel while groves have expanded from 3,000 dunams in 1948 to 20,000 dunams in 1961. The relative weight of the Jordan Valley in banana growing has declined from 75 per cent to 40 per cent in areas of groves and to 50 per cent of the total crop.

Banana growers have set up a packing centre to handle their exports in the former military cantonments near Rosh Hanikra. Choice of this location was governed by agricultural and commercial reasons. Considerable experience and know-how had been accumulated in this area, contributing to the expansion of the banana export trade. Exports which accounted for no more than 6 per cent of the crop in 1954/55 sprang to 25 per cent in 1960/61, and in volume from 865 to 10,000 tons.

But as exports grew this centre was found to be too small. It proved necessary to prolong the export season, and two other packing houses, one in the Jordan Valley to handle the early fruit and another in Hefer Valley to handle consignments at the close of the season, were established. The operations of all three, needless to say, are fully co-ordinated, in keeping with a predetermined schedule of shipments.

The Jordan Valley packing house covering 2,000 metres of floorspace (about half an acre) employs forty workers during the season and can deal with forty-five tons daily. By 1961 a total of IL. 250,000 had been invested in it.

CURING AND PACKING HOUSE FOR THE EXPORT OF VEGETABLES AND FRUIT

In 1960 work on the construction of a curing and packing house for fruit and vegetables was begun. The house was designed to handle dates in the September–December months; vegetables in December–April; and grapes in June and July. In December the date crop ends and the tomato season begins, so special arrangements were made to ensure that the handling of one did not interfere with the other.

Dates cannot be placed on the market as they come off the palm. They must be washed, disinfected, dried, matured and packed in varying forms and units. The phases require different temperatures and humidities and deep freezing. They cannot be undertaken by the individual settlement independently. Central organization and co-operation are essential.

It has become customary to employ older members of the settlements and those of limited physical capacity in the handling of the date crop. For this reason only settlers are employed. Their working

117

hours are shorter and they are provided with transport to and from the packing house.

In view of the fact that this highly complex branch of agriculture is still of comparatively small proportions it was decided to make it part of a larger enterprise to handle other fruit and vegetables. It covers 2,600 square metres, employs seventy men and women and represents an investment of IL. 400,000.

It will be seen that this plant combines diverse operations, working throughout the year in the consecutive handling of different varieties of fruit and vegetables. It is hardly necessary to point out that it has been instrumental in effecting major economies for the settlements.

THE QUARRY

Two stone quarries existed in the vicinity of Tiberias, their mutual competition enabling the settlements to buy gravel at low prices. When the possibility of a merger between the two threatened to put an end to competition and raise prices the Jordan Valley Enterprises Co. stepped in and bought a 50 per cent interest in one of them. Prices of gravel were stabilized, and a discount of 12 per cent even secured.

In the meantime, however, new problems came to the fore. The quarry had been established 25 years before, its equipment was obsolete and much manual labour was necessary. Various improvements were immediately introduced, and sales of gravel were doubled in the first year of the new partnership. Plans were prepared to increase output, to reduce the number of workers by half, to lower operation costs generally and to raise the standard of the gravel.

There are unlimited local deposits of raw material. Topographical conditions enable the transfer of the material by gravitation. Moreover extensive development work is to be undertaken and a growing demand for gravel may be expected. A further sum of IL. 350,000 is to be invested in the expansion of this enterprise.

We have reviewed the existing economic services which are steadily being developed and expanded. To complete the picture we shall outline other projects still in the planning stage but to be taken in hand in the very near future.

REGIONAL BAKERY

Nine settlements in the region have bakeries while nine others buy bread. In 1961 the price of a standard one-kilogram loaf was 24 agorot.

The regional daily consumption of bread is 3 tons. Settlements in the neighbouring Lower Galilee region requested Jordan Valley Enterprises to supply them with bread. It has been estimated that if daily consumption rises to 4·5 tons the price per loaf can be reduced to 17 agorot—that is by 30 per cent—a saving of IL. 30,000 yearly after payment of depreciation and interest. The quality of the bread will be improved while the loaf will be of full weight. The investment required will be of an order of IL. 210,000 and the bakery will employ six workers.

The competent authorities have urged the establishment of such an enterprise to serve the region.

POULTRY SLAUGHTERHOUSE

The production of table poultry by the settlements of the Jordan Valley and the neighbourhood total 1,500 tons, warranting the establishment of a regional—or more correctly, bi-regional (for the Jordan and Beit Shean Valleys)—slaughterhouse.

Other factors in favour of the development of a slaughterhouse are the difficult climate resulting in loss in weight in the marketing of live birds and high costs of transport to distant markets. According to preliminary estimates, saving on account of these two items alone would total 10 agorot per kilogram liveweight. The slaughterhouse, in which IL. 450,000 must be invested, would provide employment for forty-five men and women, most of whom would come from Tiberias.

REGIONAL FEED MIXING PLANT

In 1960 about 20,000 tons of concentrated livestock feed, valued at IL. 4 million, were consumed in this region. This figure is expected to rise to 25,000 tons annually within the next few years.

A number of enquiries into this problem have been made. The following are some of the findings:

(a) The potential consumption of concentrated livestock feed by the settlements of the region and those in the immediate neighbourhood but included in the Lower Galilee region is 35,000 tons annually. This warrants the establishment of a feed mixing plant of 12,000-ton volume, which will operate in two, and as the occasion arises, three, shifts.

(b) Already available in this region are: a number of grain silos; stores and feed mixing equipment, of which the regional plant can make use. The quality of the feed at present being prepared locally is low owing to inexperience. A central plant could secure the services of specialists and improve the quality of the feed.

(c) Most of the Jordan Valley settlements are situated within a 5-kilometre (3-mile) radius of the town of Zemah. Transport can be undertaken by the existing fleet of trucks, as a result of which substantial economies can be effected.

(d) The investment required is of a magnitude of IL. 1·5 million and the labour force—fifteen workers. Such a medium-sized feed plant could mean a saving of 6 per cent in costs of concentrates.

These three enterprises—the bakery, slaughterhouse and feed mixing plant—which are to be taken in hand in the near future—will expand the existing ramified network of economic services and bring the number of the latter up to ten.

EDUCATIONAL AND CULTURAL INSTITUTIONS

In the development of important educational and cultural institutions, Jordan Valley has served as an example to other regions in Israel.

The A. D. Gordon Institute for Agriculture and Natural Science, bearing the name of the renowned thinker who was one of the founders of Degania, comprises nine departments, as follows:

(a) A library containing 32,000 volumes.
(b) A Zoological Collection of 25,000 items.
(c) A collection of flora with 6,200 items.
(d) A geological collection including fossils and soil samples containing 1,500 items.
(e) Equipment and instruments for the use of students and visitors—microscopes, projector, micro-projector, binoculars, epidiascope, agricultural films, recordings of birds, charts for anatomy and histology and a national microfilm catalogue.
(f) An observatory equipped with a telescope.
(g) A meteorological station.
(h) A lecture hall for purposes of study, meetings, and exhibitions.
(i) The Gordon Corner.

The Institute engages in the definition of natural objects and in the study of the district, and extends bibliographical guidance.

Ohalo

This cultural instition in the Jordan Valley is owned jointly by the Executive of the General Federation of Labour (Histadrut) and the Jordan Valley region, and was established in memory of Berl Katzenelson, one of the architects of Israel society. About 2,000

120

students pass through this institution every year, the total number of days of study spent here—in seminars, etc., of varying duration—totalling 10,000–11,000. Members of the *kibbutzim* serve as lecturers, either voluntarily or at nominal rates of payment, to cover the hours of work lost by their settlements.

In addition various agricultural, scientific and artistic gatherings and conferences, as well as study days, are held in Ohalo, under the auspices of Workers' Councils in all parts of Israel. Such functions are especially arranged for the benefit of new immigrants. Activities in this category total 15,000–16,000 days per year.

The seminars differ both in respect of duration and content. Some of them—mainly for younger people—last three months: the shorter ones are for older people. In all of the activities the inhabitants of the region play an important rôle.

Beth Yerah Agricultural High School

This high school was founded in 1949. The curriculum from the outset comprehended three main trends: the humanities, natural sciences and agriculture. In its first year the school had 94 pupils; ten years later the student body had tripled to 270.

Both the teachers and the pupils experienced considerable difficulty before a suitable educational method was evolved. Primarily, as its name indicates, it is a farm school, its curriculum being based upon four years of secondary grade studies. In the first two years all subjects prescribed, including agriculture, are obligatory. In the final two the students may exercise some freedom of choice. The boys study agricultural mechanics, and the girls nutrition and handicrafts. The girls, however, if they choose can opt for zoology, biology, botany and similar subjects.

The Trades School

This school concentrates upon metalwork and agricultural mechanics, with a curriculum based upon three years of study. It is under the supervision of the Vocational Education Department of the Ministry of Labour. In 1960, 98 of its pupils came from the *kibbutzim* (67 from the Jordan Valley); 18 were from *moshavot* and 26 from the towns—142 pupils in all.

The Hashomer Hazair Educational Institute

In the Jordan Valley region there is a school set up by the Hashomer Hazair movement. This is an educational institution centring about a *Hevrat Yeladim*—Children's Group—and based upon the project method, connected with humanistic and natural scientific concentres. Most of the graduates of the school have joined *kibbutzim*.

121

The amphitheatre of the Jordan Valley was one of the first to be established in an agricultural area.

INDUSTRIAL ENTERPRISE

There are nine industrial enterprises in the Jordan Valley, one of them owned by an inter-*kibbutz* partnership of a single movement—the *Ihud Hakvutzot Vehakibbutzim*. Another factory established by one *kibbutz* is now being run as a joint enterprise by four. The remaining seven are individually owned and operated by *kibbutzim* in the region.

Thus neither the regional council nor its economic arm—the Jordan Valley Settlements—has any factories. The initiative emanated, in the majority of cases, from a few of the *kibbutzim* and within the latter from a small number of settlers.

In 1934, I recall, I was requested by a member of Kibbutz Afikim to attend a general meeting at which the fate of a small workshop established upon his initiative was to be decided. The member was technically gifted but could not find a suitable outlet in any of the agricultural branches of the settlement. The workshop in question engaged in the manufacture of small wooden boxes of the type used by dentists to hold their instruments. Over a period of one or two years it had incurred a loss of £30 (sterling) and many members of the settlement insisted that it must be liquidated.

It was a long and difficult debate at the end of which the author of this book succeeded in persuading the *kibbutz* not to shut down the workshop and to take seriously the promise of the member of the settlement responsible for it that he would expand and diversify its operations.

This insignificant little workshop that was being run at a deficit was, in a sense, the embryo of the Kelet Works, which, at one time, was one of the largest factories in the country.

It was from this factory that the initiative emanated to build a second in this district, the Sefen Works, owned and operated by a number of local *kibbutzim*. Sefen is one of the most important factories in this country.

The Kelet Works, which is still managed by the founder of that small workshop in Kibbutz Afikim, has gone through six distinct phases.

(*a*) For seven years it engaged in the manufacture of ordinary boxes, using hand tools. It was only towards the end of this period that a mechanical saw was acquired.

(*b*) Then came five years in which the works produced sewn crates, and the first experiments were made in wood-peeling for this purpose.

122

(c) A transitional period of four years during which Kelet went over to the production of plywood with very simple tools.

(d) Within the next four years Kelet embarked upon the modern manufacture of plywood, using one set of machinery.

(e) Another set of machines was introduced and three auxiliary departments were established—a saw mill, and veneer and panel sections. This phase lasted five years.

(f) In 1958 a third set of machines was installed and the development stage in the manufacture of plywood was completed.

Afikim, it should be mentioned, was launched as a purely agricultural settlement.

The importance of this factory within the *kibbutz* can be gauged by the fact that at present it contributes 77 per cent of its income notwithstanding the very considerable expansion of agricultural activities in the meantime. In 1961 Kelet's production topped IL. 10 million.

The second phase in Kelet's development set in when it reached agreement with the Allied Armies during the North African campaign, to supply wire-sewn crates for packing tins of fuel. At this time technological changes were introduced and wood-peeling equipment was installed to produce cheaper wood for the crates. The works expanded rapidly and the *kibbutz'* income from it rose steeply from 24 to 70 per cent of the total.

The fateful period in Kelet's development set in with the conclusion of the Second World War, when military orders came to an end. Kelet now embarked upon the manufacture of plywood, using available equipment but at the same time endeavouring to secure new machinery. It was necessary to wait two to three years until Europe's industry was reconstructed. Orders for equipment were placed in Sweden and the United States while the possibility of buying second-hand equipment was also considered. In 1950 the processing of timber imported from Canada, Yugoslavia and West Africa began. Use was now made of synthetic glues, the quality of the finished product improved, and attempts, which proved successful, to penetrate into the English and American markets were made. It is noteworthy that today the Kelet Works exports 60 per cent of its output and is one of the leading enterprises producing for foreign markets.

Kelet caused a storm in Afikim that did not subside for many years, and even today many of the settlers are dissatisfied with its rapid development, not for any economic reasons but because it must employ hired labour and because of the problems arising out of its close relations with business and industrial circles, both in this country and abroad.

These problems, we believe, would assume a different character if

123

the factory were situated elsewhere, not within the bounds of the *kibbutz*, and were owned and operated by a regional economic corporation.

The Sefen Works which manufactures insulating materials (celotex, masonite, etc.) belongs to a dozen Jordan Valley *kibbutzim*. It aims—with Government assistance—at replacing imported insulating board, at utilizing the by-products of the plywood factory, also local timber, and, of course, at providing an additional source of employment.

The raw materials used by this factory are: plywood waste—20 per cent; eucalyptus wood—50 per cent; pine, cypress and casuarina— 30 per cent. As a result of its consumption of local woods Sefen has stimulated afforestation in this country, and indeed the factory has planted a forest of its own. As a consumer of certain specialized chemicals it has encouraged the local manufacture of the latter.

Sefen also meets the entire local demand for insulating board, exporting considerable quantities to other countries including the United States. Its output is three million cubic metres of insulated and laminated board yearly. It is planning the production of tails-board, a porcelain substitute, as well as other products. More recently it has embarked upon the manufacture of laminated board.

Architects and builders have commended the high standards of Sefen's products, stressing their utility and the economy they allow in the use of steel, concrete, cement and labour.

Four hundred workers living in Tiberias and the vicinity are employed by the Sefen Works, besides another hundred engaged elsewhere in the marketing of its products.

Nevertheless, Sefen still has to solve a number of major problems, including the lack of expert and managerial personnel among the members of the *kibbutzim* and increasing exports, as up to the present the Works is only operating at two-thirds of its capacity.

The Eshed Canning and Concentrates Factory was founded in Ashdot Yaacov in 1937. Eshed also served as a pioneer in its field of *kibbutz* industry. In its wake two other large canneries were established in *kibbutzim*, in the central and southern districts. The three have entered into a marketing partnership.

This factory owned by two neighbouring *kibbutzim* engages in the processing of orange and grapefruit culls for the manufacture of fruit juices and concentrates, principally for export. Other products include jams, canned olives, melons, plums, cucumbers, cabbage, tomatoes, etc., which are also, in the main, shipped abroad.

The cannery suffers from fluctuating supplies of fruit and vegetables for processing. It employs 110—and sometimes more—men and women in the season.

124

Here again a major problem is the lack of specialists. Eshed, however, also suffers from inadequate working capital as large stocks of canned goods must be held over to the off-season of fruit and vegetables. The regional council extends important aid to the factory.

The Tarit (Sardinella) Kinneret Canning Company, situated in the *kibbutz* of Ein Gev, began operations in 1953. The demand for its tinned sardines, caught in Lake Kinneret, is satisfactory and no large stocks have accumulated. It holds stocks of auxiliary materials. Tnuva, the central co-operative marketing institution, acts as Tarit's selling agent.

In addition to the *kibbutz* there are two other partners in the cannery—the Kinneret Fishermen's Organization and the Ampal Corporation. The number of employees varies from thirty to seventy according to the season—in addition to a small number of members of Ein Gev and *Kibbutz* Genossar. Equipment has been improved and a deep-freeze plant installed facilitating uninterrupted production.

There are four other industrial enterprises in this region: Taglit in Ashdot Ya'acov—Hakibbutz Hameuchad, manufacturing fruit-juice extractors; an agricultural machinery factory in Ashdot Ya'acov—Ihud Hakvutzot; a mechanical metal workshop in Degania Beth, and a furniture and joinery works in Masada.

CONCLUSIONS

The foregoing indicates that at no time did the settlers of this region give any serious thought to establishing joint industrial enterprises; they certainly never considered the possibility of developing a town within their rural setting. A member of the regional council, who heads the economic enterprises of the district, told me frankly, after my lecture to chairmen of regional councils on the *agrindus* idea, that he had never thought in that direction but that his region would have to ponder the matter very deeply. When he was told that the boundaries of his region were unsuitable and should be altered, he replied that this could be done if necessary.

A cursory examination of the map of the Jordan Valley and especially of its meandering boundaries is sufficient to show that primarily the Valley and neighbouring Lower Galilee must be joined to form a single contiguous area. Geographically and economically this is feasible. (See Maps 6 and 7.)

An urban centre must be laid out. This is a matter that has not been given any thought, with the result that the economic enterprises have been concentrated in an inadequate site near the town of Zemah.

This area cannot possibly be developed into an urban centre. Unification of the two neighbouring regions must be followed by the location of a suitable site for the regional town.

Of the existing site the Report of Jordan Valley Enterprises for 1959/60[1] says as follows:

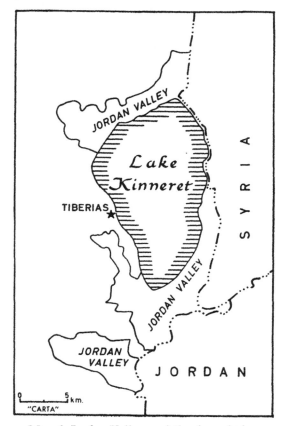

Map 6. Jordan Valley—existing boundaries

'The plans drawn up for two new factories give the compound the character of an industrial zone. It is small, however, no more than 50–60 dunams in extent, half of which is occupied. We don't waste a foot of land and build very closely. We have waged a struggle with external planning institutions and also with the regional authorities. In 1961 we shall have to find room for other enterprises which will

[1] 'Enterprises—Zemah', Jordan Valley, p. 2.

have to take over land at present at the disposal of experiments. It is highly regrettable that the enterprises must be set up at the expense of the very beneficial work being done in the experimental field. The leaders of the community, however, are to blame for not having earmarked an extensive area for public purposes. Necessity, of course, knows no law.'

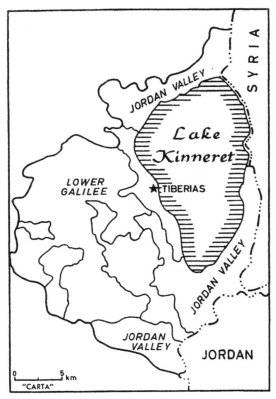

Map 7. Jordan Valley—proposed boundaries

In addition to the enterprises which we have described already in operation or under construction, a number are being planned and will be set up in the not distant future. These include a regional grain silo, a butchery, a regional laundry, a filling and greasing station, a citrus packing house, a water society, a drainage authority, etc.

The author of the section of the report quoted is fully aware of the mistakes that have been made, the short-sightedness and lack of

127

vision among the region's leaders and the remedies that are possible. He sums up in the following terms.

'The achievements registered mount up to a very gratifying score, but the problems and obstacles still confronting us are hardly of smaller dimensions. There is much room for improvement, many gaps to be stopped. The dangers menacing public enterprise grow together with the latter, and a constant and unremitting effort is necessary to ensure efficient and economical maintenance and operation. Here the old saw that all beginnings are hard is not apt. The contrary is true. It is easy enough to launch an enterprise. It is a difficult and onerous task to continue. The rapid pace of expansion now necessitates development in depth, the safeguarding of the foundations. It is only in this way that we can ensure further expansion.

We need public sympathy, public counsel, public authority and a favourable public opinion. And just as we have benefited from these in the past, we trust that they will be forthcoming in the future, no less.'

This is a sound, sincere and sensible statement but it is wanting in perspective. Another project, another field of activity, a few thousand more square metres for that restricted zone near Zemah? No!

After this chapter had already been completed I was invited to visit the Jordan Valley. All leading settlers of the region took part in the tour that was arranged and in the discussion with which it concluded. The Jordan Valley is bountifully endowed with initiative and with managerial talent. Those present agreed that it was not enough to set up additional enterprises. Today the latter are scattered over the entire Valley—the services here, the Sefen Works there, the educational and cultural institutions anywhere. In this region inter-*kibbutz* co-operation for the joint operation of branches of agriculture is not being considered; but there is a definite prospect of co-operation between the settlements in economic and cultural enterprises. Indeed there was no opposition to the suggestion made that the Jordan Valley should link up with the neighbouring regional council of Lower Galilee. The self-imposed isolation of three settlements within the region—in the form of local councils, enjoying all the benefits of regional organization but bearing none of the burdens— was irksome to the leaders of the area. The idea of a co-operative town in the centre of the region is still alien to them. The prospect of establishing useful projects has been studied but not the need for a central site for them. We must hope that they will come to appreciate this need.

The following suggestions could be usefully implemented in this important region.

128

(1) The Jordan Valley and Lower Galilee should be united to form a single region comprising more than thirty settlements, doubling the population.

(2) A regional town should be laid out with the least possible delay in keeping with the principles of the Location Theory.[1] The site should allow adequate room for expansion to meet the needs of future generations. In this town all services, factories, cultural and municipal services must be established.

(3) The regional town must be developed to form an integral extension of the agricultural settlements, socially, economically, culturally and administratively. This town must be constructed on fully co-operative principles by all the settlements in the region; together, the town and the settlements will constitute a single unit— the Jordan Valley *agrindus*. In the same way, any valley or any other tract of land anywhere in the world, where suitable economic and social conditions exist, can constitute an *agrindus*.

[1] J. H. von Thünen (1783–1850), in his *Der Isolierte Staat in Beziehung auf Landwirtschaft und National Ekonomie* (1826), developed a theory according to which the territorial distribution of agricultural production is determined by prices of the products at a consumption centre and the transportation costs from the locality of production. To illustrate his ideas he constructed a hypothetical 'isolated State', in which the great city is encircled in a series of belts in a given pattern of different farm branches. Thünen was aware that a differential rent results from advantages of location and from differences in soil quality. This theory, which for a century attracted little attention, has served in our days as the basis for a new discipline of 'Economics of Location', which has been elaborated in a series of new studies, e.g. in those of August Lösch, who developed a theory of economic regions and presented an equilibrium system describing the inter-relationship of all locations. Other works dealing with this field have been written by E. M. Hoover (e.g. *The Location of Economic Activity*), Walter Isard in several publications and especially in his 'Location and Space-Economy', a General Theory Relating to Industrial Location, Market Areas, Land Use, Trade and Urban Structure, and others. Many works have been devoted to Industrial Location.

XI

SHA'AR HANEGEV

AT the lecture—already referred to—given to leading members of the regional councils, when I stated that it was not unfeasible that at some time in the future the regions comprising only *kibbutzim* could set up 'a commune of communes', the chairman of the Sha'ar Hanegev regional council interjected, 'We have a commune of communes already.' In the discussion which ensued many of those present remarked: 'Small wonder. You have a homogeneous region.'

But these remarks do not reflect the true situation. It is correct that the Sha'ar Hanegev region is fundamentally different from others; nevertheless it has not achieved regional unity and is not even distinguished by regional co-operation. To call it 'a commune of communes' accordingly is unwarranted under the circumstances.

It is not even completely correct to say that it is homogeneous. The region, it is true, has nine *kibbutzim*, all affiliated to the Ihud Hakvutzot Vehakibbutzim, but it also has one *kibbutz* of Hashomer Hazair —Ruhama, and a *moshav*—Yachini. The two estates of Ibim and Marit, of course, do not constitute a problem. The town of Sderot, however, though located within the region, does not belong to the regional council, but is a municipal enclave—it is constituted as a local council—and even in greater measure an economic and social enclave.

But perhaps in this district it would prove easier to attain our objective. The difficulties that must be surmounted however are the size of the region, comprising no more than eleven smallish settlements, and the town. It is also possible that a young and dynamic area like this will chart its own course.

We shall describe developments in this region in recent years and then outline and discuss views on its future development.

130

SHA'AR HANEGEV

MUNICIPAL ACTIVITY

The council has engaged in planning, reconstruction and the paving of roads and pavements. It provides transport for children to and from the schools and kindergartens, and for bringing provisions to the dining room in Yachini. It maintains a fleet of three ambulances, each in a different point of the district, at the disposal of pregnant women, the sick and injured, and doctors in cases of emergency. The council's other spheres are sport and recreation, fire-fighting, sewerage, the construction of air-raid shelters, dams, etc. The pride of this region, which as stated is not overlarge, is the amphitheatre which can seat an audience of 2,000.

The sanitation work of the council includes the extermination of rodents, flies, mosquitoes and other pests.

EDUCATION AND CULTURAL INSTITUTIONS

The region has a primary, secondary, vocational and evening (for adults) school. The Sha'ar Hanegev Regional School, equipped with post-primary continuation classes, is administered by the regional council and the local settlements of the Ihud Hakvutzot Vehakibbutzim. It is based upon the educational principles of the latter, but as a state school is open to children from all settlements in and outside the region. Each of the settlements contributes towards the upkeep of the school according to the number of its children attending it. Every settlement, moreover, is under an obligation to train and to provide teachers and other personnel in accordance with the same principle. The regional council contributes towards the cost of training teachers, who must be approved by an Examining Committee before taking up their positions. The school council comprises the chairman of the regional council, the principal of the school and two representatives of each settlement. The Statutes define the competence of the principal, the executive committee, the pedagogic council, etc. Plans for 1962 include the construction of new classrooms, a manual training centre, swimming pools, clinics, roads and a cultural centre in Yachini, all of which, together with the equipment required, will call for a sum of IL. 515,000.

SERVICE BRANCHES

These include a regional garage, employing twenty-five workers, a filling station, marketing facilities for poultry, lucerne and dried fruit, tanks for cooling milk, agricultural equipment and machinery, radio and the maintenance of an office in Tel Aviv.

131

The Sha'ar Hanegev region is planning the construction of a regional laundry, the establishment of a purchasing agency and a joint transport service.

PRODUCTIVE ENTERPRISES

The region undertakes agricultural operations which are beyond the capacity of any single settlement. For this purpose the region organized the Talmei Hanegev Society which in 1961 harvested the beet crop on an area of 2,500 dunams, picked the cotton crop of 5,000 dunams, scattered stable manure over 8,000 dunams and liquid fertilizer over 30,000, and also planted lucerne and gathered the potato crop. The council has a shovel-dozer for earth-moving jobs, undertakes the grading and cold storage of apples and the dehydration and refrigeration of vegetables. The council is planning the construction of a 400-ton cold-storage plant.

In all these operations the council observed the principle of mutual responsibility, even before incorporating this principle legally in regional co-operation. In the summer of 1961 the council harvested all the sugar-beet sown in the district. Priorities were fixed on the basis of the size of the beet-fields and the date of sowing. It had previously been agreed that the priorities would be adhered to, but when it was found that the beet of one *kibbutz* was stricken by some disease and that if the pulling was delayed severe loss would be incurred, it was resolved to harvest the entire crop at once, and not in three rounds as is usually the case. Of course some loss was suffered by all other settlements as a result.

Another example: ten clinics are being erected in the region in keeping with the funds available. One settlement is using only concrete in the construction of its clinic. The extra cost is considerable and the region would have been fully justified in imposing it upon the settlement in question. But this was not done.

And finally a third example: one settlement needed a harvester, but did not have the money to finance the purchase. Though none of the other settlements required one, the former requested the council to buy such a harvester for its inventory of heavy equipment, undertaking that it would not remain idle throughout the season, thereby ensuring that it would pay for itself. The council decided to buy the harvester.

These and many other instances not cited serve to indicate that collaboration has already assumed many and varied forms, though in theory no principle of regional co-operation has been officially adopted. Joint services and *ad hoc* partnerships have been set up. Regional leaders, for example, are contemplating the possibility of

132

joint cultivation of field crops. Of course, combined harvesters and other equipment must be purchased for this purpose.

INDUSTRIAL ENTERPRISES

The settlements have not gone in for industrial development, and indeed there are only three factories in the region: a metal products plant in Dorot; a brush factory in Ruhama and a dehydrated vegetables factory in Bror-Hayil.

Dorot manufactures taps and specially cast bronze irrigation cocks, for which there is a good demand in Israel. Sixty workers—twenty of them members of the settlement—are employed. The brush factory at Ruhama works for both the domestic market and for export, and employs some dozens of workers. Bror Hayil's dehydration plant, in which Swiss investors hold an interest, is of far more ambitious proportions. Only 20 of its 120 workers are members of the *kibbutz*. The factory engages, both for the local and for foreign markets, in bulk production and packing, the dried vegetables being sold in large metal casks. Packing in smaller quantities is undertaken by wholesale merchants. The main problem of this undertaking is that it has not yet reached optimal production, which would ensure profitability. It can process forty tons of vegetables daily, which must be supplied by the settlements in the neighbourhood. Unfortunately, however, supplies are not regular, and for most of the year it averages only thirty tons.

The managers and experts employed by all three plants have acquired considerable technological experience and ability. However, in the largest of them the impression gained is that its dimensions have already exceeded the capacity of any single settlement and that if the region were willing Bror Hayil would not oppose its conversion into a regional enterprise.

The manufacture of taps and brushes is of a purely industrial character, utilizing outside raw materials; dehydration of vegetables, however, is definitely of an agro-industrial character, deriving its raw materials from the immediate neighbourhood. The latter, accordingly, has a direct interest in its successful operation, for if it should grow in the future, as may be expected, it may become burdensome for the settlement that founded it.

The regional council operates three plants—a lucerne drying mill, a cold storage and freezing plant, and a cotton gin.

The lucerne mill was erected in 1955 with an annual capacity of 3,500 tons. High yields have been registered in this area and output of the mill has averaged two tons of meal per dunam. The settlements are given instruction in the proper cultivation of their lucerne

133

fields. A sum of IL. 400,000 has been invested in the mill, which employs 27 workers and has an annual output worth IL. 750,000.

The *Of Kar* ice factory, slaughterhouse and deep freezing plant was established in 1958. It operates for 220 days a year processing 350 tons of dressed poultry. (In fact this figure has been surpassed with the same staff and equipment.) The labour force is 90 men and women, the investment IL. 700,000 and the annual output IL. 3 million.

The *Cotton Gin* is at present under construction. The investment required is IL. 1 million. Not all of this sum has so far been raised.

In addition to its investments in the municipal field the council is planning to sink a sum of almost IL. 2 million in the cotton gin, the cold storage plant and slaughterhouse, Talmei Hanegev and the regional laundry.

The regional school and the area for development are situated not far from the town of Sderot. The development area, however, is too small to meet the needs of an *agrindus*. Of course we have already stressed the special, largely homogeneous character of this region. In the long run, however, it is doubtful whether eleven settlements can sustain an *agrindus* of optimal proportions. It might be advisable for Sha'ar Hanegev to coalesce with the two adjacent regions to the south. The combined area would then have more than thirty settlements and would be capable of developing as an *agrindus* with an important urban centre—perhaps the same town of Sderot, but upon a co-operative basis and integrated with the region. There is the alternative of joining up with the Ascalon Coast region with sixteen settlements, though it will prove far more difficult to integrate the town of Ascalon, which has developed rapidly in recent years into a medium-sized centre. Alone Sha'ar Hanegev is capable of developing forms of co-operation that are uncommon in this country. It is doubtful whether this course will remain open to it if it links up with some other region to the north or the south. It will have to decide which it prefers—quality or size.

Some time ago Sha'ar Hanegev arranged a symposium of a unique character, in which representatives of the local settlements and leading members of the Ihud Hakvutzot Vehakibbutzim took part. A brief outline of the discussions, we feel, will contribute towards clarification of the thesis we are developing.

The object of the discussion within a circle of about sixty men and women was to develop ideas on the subject in the settlements. It seems that local leaders in the course of their daily work had discovered that the new regional unit, called the Council, was—to use their own words—'a factor to be reckoned with which, however, we have

134

not yet learnt to use properly'. The opening speaker set the keynote when he asked:

'Is this the course we must follow? Should we continue to make our plans upon the basis of partisan demands addressed to one settlement or another? Up to the present [he pointed out] we have not yet formulated any regional statutes, any carefully thought-out and planned course of action for what is being established in the region. But another, highly significant attitude has manifested itself—that of the various official bodies, the settlement institutions, the Ministries of Agriculture, and of Trade and Industry, which obviously regard the regional council as a serious and effective body. Generally speaking throughout the country regional councils are undertaking important functions and projects. Ministries address themselves to us, various enterprises and activities are suggested and even money is offered. This is very encouraging. We are very circumspect . . . for we are conscious of the abyss between the course local leaders would like to adopt and the reaction in the region, in the settlements . . . The obligations of the latter towards the centre are of a very tenuous character . . . In fact the settlements have invested no more than 3 per cent of the value of the assets of the region . . . If local leaders are not conscious that they are being supported, if they do not feel they have ground beneath their feet, it is difficult for them to act . . . Manpower is the major problem we have to cope with . . . Every new project undertaken requires the enlistment of people from the settlements, capable of filling central positions . . . and it is not easy for the settlements to forgo the services of members of this calibre . . . Indeed in the majority of cases the response is in the negative . . . On the other hand the situation is paradoxical. Let us take for example one area in our own region, where there are 4 kibbutzim cultivating an unbroken stretch of 12,000 dunams of irrigated land. There are 4 irrigation managers, 4 cotton managers, 4 sugar-beet managers, 4 tractor managers, 4 farm managers running to the same institutions and waiting in the selfsame queues. I could also speak about fragmentation of holdings, about wastage in use of tractors, but we are told: "We are short of manpower".'

The problem put to the settlements was that of regional co-operation or even regional unification. About these two issues a lively debate is at present being waged in the settlements. In opening the symposium the chairman of the regional council declared:

'Regional co-operation is already a fact, though we are not aware of it because we have not put it all down on paper, and tried to see the impact of the region upon the settlement, and the degree to which the settlement is dependent upon the region. I have made a list of the things we are doing and I have come to the conclusion that if we add only one or two more the dream we call regional co-operation is not so far from realization.'

'. . . The council operates a lucerne meal mill, a laboratory for field services; we have equipment and tanks for cooling milk for the settlements; we have organized anti-pest and anti-rodent measures; we harvest all the sugar-beet; we pick all the cotton; scatter all the stable manure and liquid fertilizer; we pull the potatoes with combined harvesters; dig *kurkar* and scatter it in the settlements; market all the poultry; grade and cold-store all the apples. We have an office in Tel Aviv where the treasurers and buyers meet and from which contacts with the enterprises are maintained. We have a wireless telephone, a regional school and an amphitheatre. We are conducting experiments in fruit drying; maintaining the internal services of the projects; operating a garage and filling station. We shall soon set up a cotton-gin, a regional laundry, a slaughterhouse and a cold-storage plant. This, of course, is all in addition to normal municipal services . . . Joint operations, in themselves, have opened up new vistas . . . We are getting used to thinking on new lines . . . We have handled and managed enterprises amateurishly, and if we have succeeded it is because of our strenuous efforts, because of the incalculable amount of overtime we have invested and the dedicated work of our members . . . We have not studied the subject; we have not undergone any vocational training. Our success must be attributed to our own devotion to an idea. Let us send members for training and study and we shall achieve an even greater measure of success . . .'

'In our talks we have often contemplated regional co-operation as a very distant prospect, an ideal which one day we may realize. But I do not know whether regional co-operation has not become a vital need . . . Some people are thinking in terms of removing all frameworks, of leaving only the embracing framework of regional co-operation, of developing the region as a unit. The psychological revolution that is necessary for the transition from individual to collective living is not greater than that from collective living in the *kibbutz* to regional co-operation. These problems have been studied by a group of 50–60 people, but 1,900 other people in the region have given no thought to the matter; they are not clear about what is being done in the region, in the council, in the joint enterprises . . .'

The speaker stated he was not sure that this idea could be put into practice in the *kibbutz* movement as a whole; in the Sha'ar Hanegev region, however, favourable conditions existed 'to create something that while not new in conception has not been put into effect in Israel'.

Some of the participants expressed their reservations. Others envisaged the expansion of the existing *kibbutz* framework to incorporate complete co-operation in all fields. But there were also those who feared that over-expansion would weaken co-operation within the *kibbutz* as a unit, while regional co-operation might develop into an instrument of administration and no more. There were even some opposed to closer co-operation in individual farm branches which they said constituted the very basis of *kibbutz* democracy, of education towards mutual responsibility, of comradeship. If the essence of comradeship should assume the form of administrative dependence, comradeship upon which *kibbutz* co-operation is grounded would lose all meaning. Thus, it was argued, regional co-operation must be based upon local co-operation. Economic, cultural and organizational services which are beyond the capacity of a single *kibbutz*, whatever its size, must come within the field of regional co-operation. Efforts must be made in this direction to ensure that the *kibbutz* movement is not left out of the process of industrialization of Israel. This is the specific task of the region, but local initiative must flow along the social channels, marked out by the *kibbutz*.

But however this process may be regarded from a practical point of view, it is also of major significance as a matter of social principle. Economic collaboration, it was emphasized, does not carry with it collective collaboration. Partnerships may be very successfully developed in the region and at the same time the dykes holding back the flood of hired labour destroyed, for a considerable number of *kibbutz* members may be converted into managers of thriving economic enterprises. Even joint operation of the regional amphitheatre can assume an economic character, divorced of the principles and values upon which *kibbutz* society is based. If regional co-operation is of a *kibbutz* character it must assume responsibility in many spheres: economic, social, educational, ideological, as well as in security. The question is whether this responsibility is qualified or not. There are factors which have a retarding effect and others which may act as stimulants. Even the maximalists are of the opinion that the advance in this sphere must be phased, though it must have a definite goal. Regional co-operation was presented as an ideological category, towards which action must constantly strive. The common fund with which the G'dud Ha'avoda experimented and failed represents an idea; it could be tried again to test if it is not feasible today. It is not necessary at the present time for the economic unit to be identical with the social unit. A *kibbutz* in the region may preserve its identity as a

137

social unit and at the same time it can be, from the economic aspect, a constituent member of the region of Sha'ar Hanegev. But even the advocates of full co-operation in the future are in favour of maintaining the identity of society and the economic settlement unit and of developing, in the highest degree, mutual aid between the settlements in the economic, social and cultural fields. But it is not necessary, at the present juncture, to elaborate a final goal. In the phased advance regional co-operation, principally of an economic character, must be regarded as a preliminary stage. For the present, a suitable atmosphere must be created, in which a course of action can be charted. 'In the time of abysmal decline we live in, it would be just as well if we were capable of running towards something even if that something is not quite clear,' it was said. When one of those who took part in the debate was asked why he had spoken in such a contradictory fashion, for and against, he replied: 'That is how one climbs a mountain. You cannot go straight up. Your path necessarily twists and turns.'

Some of the participants insisted that the realities of the situation must be seen clearly. The basic and far-reaching differences between collectivism, as practised in the *kibbutz*, co-operation of a more restricted character and ordinary partnership should be underlined. Each of these socio-economic forms has its contribution to make in promoting the interests of the settlements and the region. Within twenty years, perhaps, the question may assume a different form. Today it is necessary to build in keeping with existing concepts and experience. The regional economy possesses extensive room for development. There is no reason why the individual settlement should fall behind. It is not necessary in every case to lay down a joint framework. If some course is charted it must be pursued, developed, expanded. Thus branches of agriculture, productive enterprise which rounds out the economy, operations which are beyond the capacity of a single settlement, can be developed. At all events the consensus of opinion is that industry should be developed on a joint basis. This is also the case in regard to cultural and educational institutions. The most irksome problem of the *kibbutzim* is that of hired labour. If regional co-operation could offer a satisfactory solution it would win the support of all interested parties. But we shall return to this question of hired labour.

Someone at the symposium quipped: 'It seems that we achieved regional co-operation on one point—fear of regional co-operation.' The speaker regarded regional co-operation as an ideal, but in his view, it is hard indeed for any man to revolutionize his life twice. A person who has adopted collectivism cannot possibly take the further step of collectivizing the collectives. That is a task for the next generation.

138

Kibbutz members, from other regions, who participated in the discussion, also voiced diverse and often contrasting opinions. They operate on a nation-wide scale on behalf of their movements. They found that Sha'ar Hanegev is not different in its co-operative enterprises and services from any other region. New developments in agriculture make it necessary, in most areas, to establish enterprises, mainly in order to solve difficult technical problems.

It was pointed out that in old-established settlements it was more difficult to achieve some degree of regional co-operation. In the younger settlements, however, thinking upon these lines was more developed.

Many roads lead to regional co-operation. Joint ventures in various branches are possible without co-operative enterprises having to be set up. The converse is also true. Local conditions and development are an influential factor. Thus, if in this region the *kibbutzim* are accustomed to the idea of pulling their beet on a joint basis, they have not far to travel to arrive at joint ploughing, sowing, etc. It is quite feasible to establish compensation and equalization funds, which indeed already operate on a national basis in the form of mutual insurance, marketing, etc., without setting up a common account. Here or elsewhere such a common account may be set up in the course of time, but it is by no means an indispensable precondition for regional co-operation. Forty and even thirty years ago the optimal size of the *kibbutz* was the subject of an exhaustive discussion. One extreme view—adopted by some *kibbutzim*—was that the *kibbutz* should not comprise more than thirty units. Degania Aleph, the first *kibbutz*, split up when it exceeded this figure. Today that controversy has been relegated to limbo and those who once so stoutly defended the small *kvutza* today speak of 300 units as the minimum ensuring the stability of the collective, and allowing it to achieve the cultural level that contemporary men and women insist upon. But then many other issues which were hotly contested years ago have been forgotten as conditions have changed. In our own day people have more spacious ideas. Fifty years of experience have proved that the smaller the unit the more difficult conditions are. It is true that there are only two *kibbutzim* in Israel with 300 families, but already 500 is considered practicable in the near future, and one or two thousand at a more distant date. Even this goal falls short of the population of a small town. The technology of big machinery presupposes larger economic units. The towns that have sprung up under the impact of special circumstances will change as the younger generation grow to manhood; indeed it would be easier then to integrate the town and its population within a framework of regional co-operation.

The *kibbutz* was not the product of planned development. Many of those who joined the collective settlements did not have any clear idea what shape the structure they were building would eventually take. This may be the case with regional co-operation, too. It is a problem all regions will have to cope with, but Sha'ar Hanegev offers a convenient object lesson, just as it has learnt and must learn from others.

A *kibbutz* settler from another district, who was one of the pioneer thinkers and builders of the large collective settlement, was deeply gratified by the ferment of ideas he found in the Sha'ar Hanegev, which he is convinced will continue until the dream he has cherished for forty years is consummated. He related a story of an American tourist, an economist by profession and a socialist by conviction, who, when he had been shown all that the *kibbutz* had achieved, declared: 'All this is interesting, but it does not interest me until you show me how to make a *kibbutz* of New York ' To this the settler replied: 'New York can be converted into a *kibbutz*, but that is not my business. What I am concerned about is how to link up with the neighbouring *kibbutz* in my own lifetime. The next generation will tackle the problem of New York.' In regard to Sha'ar Hanegev he said: 'If there are so many settlers who have set their hearts upon unification, let us encourage them and let us learn from them how to set about the task. Their experience will teach us how to continue, for we must never desist. Man is restless, he is constantly seeking.'

Collaboration is a term that can give rise to error, and those who took part in the debate defined it as they understood it. Collaboration exists in a partnership between two people; it can also be expressed in a co-operative or a commune. Most of the participants envisaged a form of regional co-operative, though there were also those who spoke in terms of a regional commune, which, as yet, is non-existent.

If only we, in our generation, could establish co-operation between neighbouring settlements on a regional basis and thereby by the creation of joint enterprises assist the individual settlement in the economic, social, cultural, educational and organizational spheres, we should be content. Such co-operation would be of immense immediate importance as even the largest settlement cannot independently meet important needs, while the accelerated pace of development calls for new dimensions. Regional co-operation can meet these demands provided it is not only of an economic and technical character but is sustained by the principles of mutual aid and responsibility.

Thus, should the Sha'ar Hanegev settlements really be interested in creating such a unique social entity, let us encourage them, let us not thwart their efforts by considerations of municipal convenience

140

or economies of scale, even though eleven villages may not be able to develop an industrial centre of adequate proportions. If they should choose to embark upon regional co-operation, let us advise these two neighbouring regional councils to unite, to form a single region with thirty-three settlements which is a more suitable unit for an *agrindus*.

But there are other alternatives, too. Sha'ar Hanegev can join up with Ascalon Coast, with its eighteen settlements, to the north. It may be more convenient for it to take Hevel Ma'on, with its ten settlements, all *kibbutzim*, to the west as its partner. An objective observer looking at Map 1 may be inclined to reorganize the six regional councils—Ascalon Coast (39), Sha'ar Hanegev (42), Shelahim (44), Azata (43), Merhavim (46), and Hevel Ma'on (48)—into two councils and two *agrindi*.

In the summer of 1961 I was invited to address leading settlers in the Sha'ar Hanegev region on regional co-operation. Many questions were put, but it seemed that finally both the lecturer and his audience reached a considerable degree of agreement. The principal—and most difficult—question was how to solve the lack of suitable personnel. The only, though admittedly inadequate, reply that could be given was that thirty settlements joining their forces offered a better prospect of finding leaders of the requisite calibre than ten settlements. There are laws, it seems, governing the optimal size of *agrindus*.

XII

THE LACHISH DISTRICT

W E have included a discussion of the Lachish District in the present study for two reasons: (a) the work of settlement was launched only a few years ago, viz. in 1955, and proceeded according to a special plan; (b) this is an area with settlements only of the *moshav* type. Of the twelve villages here ten are ordinary co-operative *moshavim* while the remaining two are *moshavim shitufiim*.

Lachish is a new development area in the southern part of Israel, extending from the Judean highlands in the east to the coastal plain in the west. Its altitude ranges between 100 and 150 metres above sea level. The area is traversed by many water courses, draining into the Lachish Brook, which gives the region its name. These tributaries are dry except in the rainy season, and there is only water all the year round in the brook for hardly more than a mile from the sea. Of the 900,000 dunams of the District less than one-third—lying in the west—is fertile. The eastern section, at the foot of the Judean Mountains and the Jordan border, is an eroded highland, only part of which is suitable even for grazing.

Lacking local water resources the district remained undeveloped and unsettled until only six years ago, when a pipeline leading water from the Yarkon River thirty-seven miles away was constructed.

A large team of planners, technicians, instructors and administrators, put to work by the Israel settlement authorities, produced a detailed project envisaging the settlement in Lachish of 36,000 persons, half of whom would support themselves by agriculture, and the balance would engage in services and in industries to be developed in the town of Kiryat Gath. The town was laid out on a range of hills, about forty miles from Tel Aviv to the north, with Jerusalem to the east and Beersheba to the south. It was planned for 14,000 inhabitants.

142

Originally the economic basis of the district was envisaged as entirely agricultural. The main branches of farming were to be industrial crops—cotton, oil-seeds, ground-nuts and sugar-beet—as well as grain, and sheep and cattle raising. Today Lachish makes a respectable contribution towards the country's cotton and beet-growing. From a dry-farming area it is now an area of intensive irrigated cultivation. In the past hundreds of dunams were necessary for a farm unit; today holdings are 40 dunams (10 acres) in extent, of which 25 are under field crops, 5 orchards, and the balance earmarked for fodder.

We have already pointed to the fact that the settlements in Lachish are exclusively of the *moshav* type. The problems of industrialization, accordingly, assume a different character, and for that reason we shall describe their farming operations. A holding is based upon the early investment of 400 days of work, by the father of the family and one of his sons. Another standard of calculation is the soil and water unit, requiring ten days of labour for every dunam. Part of the produce of the farms is sold to the population of Kiryat Gath, but the *moshavim* raise mainly agricultural raw materials which must be processed in factories. This latter function is reserved for the region's urban centre. So it is in Kiryat Gath that the agricultural produce is worked up, where storage, marketing, transport and service facilities are concentrated.

In areas where most of the settlements are *kibbutzim* the decision to establish urban centres with regional functions must be taken by the villages. Their large economies constitute, in the majority of cases, integrated units supplying services and maintaining industries, in addition to their farming operations. Indeed they leave little scope for the urban centre. However, in an area settled by *moshavim*, like the Lachish District, a rural centre is necessary in addition to the urban centre, to extend the social and economic services which the single village is not capable of providing. The reason, of course, is because the *moshav* is designed to engage solely in agriculture. The services, supplies, marketing and processing facilities must come from some centre near by. This, perhaps, explains the rapid development of Kiryat Gath.

Most of the farmers in the Lachish District were settled in the villages set up in 1955–56, under the Ship to Village scheme. The newcomers were immigrants to whom agriculture had been completely alien in the countries from which they came. They are accustomed to a patriarchal type of society and their capacity for co-operation is very restricted.

In order to endow the new settlers with a feeling of good neighbourliness, of security, of social ties, the villages were planned—

143

especially in this District—for no more than eighty to a hundred families. Care was also taken that the members of the *moshav* should come from the same country of origin. Attempts to settle members of different Jewish communities in the same village proved unsuccessful. As a result of this rule a number of villages in the Lachish District settled by newcomers from Asia or Africa have no more than 40–60 families. But this system has, to a large extent, ensured the social and moral stability of the new villages.

The villages are under the watchful eye of a team of instructors who come under the authority of the Jewish Agency Settlement Department's Regional Centre. The head of the team guides the settlers in social and administrative affairs, organizes local institutions as well as its relations with external bodies. The team also includes two instructors in farming, one in home economics, education, cookery, health, etc., and a nurse. In the transitional period the instructors live in the village. Most of them come from established *moshavim* and *kibbutzim*.

The settlers take up the burden of their farms step by step. First they are employed on improvement of the soil and as building labourers at a daily wage by the settlement authority. In this initial period the farm is managed by a representative of the authority, the immigrant being regarded as a candidate for settlement. There were, indeed, not a few candidates who failed to pass the test of this probationary period and did not settle in the village. The farm is transferred to the settler after the passage of a number of years.

In the Lachish District an interesting experiment was made. A Rural Centre was established to serve clusters of 5–6 *moshavim*. Regional planning is based upon three planes. At the base is the farming village, then comes the rural centre and finally, at the apex, the urban centre.

The farming village has its social, organizational, economic and educational services to meet immediate needs. These include: the secretariat, kindergarten, club, clinic, armoury and synagogue.

The rural centre serves five or six villages in the immediate neighbourhood and is the location of certain larger services such as the primary and high schools, evening classes for juveniles and adults, a club for the youth, a village hall where cinema and other shows can be staged, a library, a clinic where a full-time doctor is in attendance and which medical specialists visit regularly, a Red Magen David station, social welfare centre, co-operative store, tractor station, store for the products of the workshops (two metal and one carpentry) and the offices of the council. The inhabitants of the rural centre are the personnel of all of these services. They cultivate allotments near their homes.

In view of the homogeneous character of the *moshavim* the rural centre plays a highly positive rôle in promoting the integration of immigrants from diverse countries to form a single Israel nation. The success of this method is already apparent to all inhabitants of the region. There is, of course, the danger that a social gap may keep the inhabitants of the centre and the villages apart. The former, after all, are specialists, managers, teachers, etc., mainly of European origin; the villagers come from backward countries. New ideas have been put forward to forestall undesirable developments. One of these is to distribute the teachers in the various villages. Another is to establish small factories in the rural centres, where the villagers—or rather mainly the younger element—will find employment, thereby changing the communal and social composition of the former. The integration of established residents and new immigrants will be put into effect gradually and naturally, largely through the agency of the common services, such as the clinic, the sports ground, the tractor station, etc. But the principal instrument, it is hardly necessary to point out, will be the school. Here the children are educated, in the widest connotation of the term. They do not merely imbibe knowledge; they are taught how to dress and how to eat, the elements of personal hygiene and of good manners. The schools and the club serve as the channel through which these positive influences are transmitted into their homes and to their parents.

The urban centre is the seat of the provincial administration and the local centre of commerce and industry. Kiryat Gath, founded in 1956, bears the name of the ancient city of the Philistines in this region. The town has a heterogeneous immigrant population and a diversified industrial complex has developed.

The foundations of the town were laid on 11 January 1956, when twenty-three immigrant families came here immediately after disembarking at the port of Haifa. Every family was allocated a two-room flat. The newcomers were welcomed by representatives of the Ascalon branch of the Working Mothers Organization of the Lachish District, the Jewish Agency's Absorption Department and the regional council. They were examined first by the social welfare officer and then by the doctor, both of whom gave them a clean bill of health. The regional council provided prayer books and a Sefer Torah, welcoming the first settlers, who had come from the Atlas mountains in Morocco, with the benediction, 'May you come in peace to rest and your heritage.' Ten days later the cornerstone of a cotton spinning mill was laid. This was the second industrial enterprise in the town. The first, a cotton gin, processing the crop of the southern district, had been established four months before.

On the day after the arrival of the immigrants a regional burial ground was consecrated and the construction of the fence and gates begun. Two air-raid shelters were also built.

Five years later, at the end of 1960, Kiryat Gath had thirty-six workshops and factories, with another seven under construction, eight in the blueprint, and six in the preliminary planning, stage. The twenty-eight workshops for metalwork, joinery and furniture, fashion goods, foodstuffs, etc., employed another 123 workers, and would, when fully developed, employ 218 in all. They represented a total investment of IL. 500,000. There were also eight factories— spinning, weaving and knitting mills, a cotton gin, cold storage plant, chemical factory and bakery. These represented an investment of IL. 8 million and employed 494 workers. When in full production they would employ 1,104. Since that date the construction of seven other plants has been completed. An aggregate of IL. 17 million has been invested in a sugar refinery and another million in a textile mill.

Other enterprises being projected include a large textile plant and another spinning mill, in which IL. 8 million will be invested. The two will employ 650 hands.

In fixing the location of Kiryat Gath planners were guided by the fact that the site was at the junction of the main roads traversing the Lachish District. In the west, four rural centres each serving 4–6 villages form a semi-circle. The town is on an average no more than about six miles from the centres and suitable transport can be developed.

Town plans provide for three residential areas, in keeping with topographical conditions. Set within these areas is a commercial, transport, administrative and cultural centre. Near the railway is an extensive industrial zone and spacious sports fields which serve the region. All three residential areas have already been built up. The factories, stores and sheds of the industrial zone already make up the characteristic feature of the new town. The modest buildings of the Settlement Department and the Development Authority, set within trees and flowering gardens, constitute a landmark. Near by is the commercial centre, the bus station, the cinema and the archaeo- logical laboratory built around a small municipal garden.

Originally, when the population was planned for 6,000 persons, semi-detached houses were built, each unit occupying 750 sq. metres. When plans were altered to provide for a population of 14,000, long terraced cottage buildings and multi-apartment structures were erected, with the result that the town acquired a rather more sophisti- cated urban aspect. These buildings have been constructed by the Government's Housing Authority. But 20 per cent of the houses will be built by the residents themselves, and there are already symptoms that the town will exceed the bounds planned for it. In October 1960 2,104 families and 120 single men and women, that is 9,333 persons, lived in Kiryat Gath.

In 1959 an economic survey of the region was prepared at the invitation of the Ministry of Labour. The prospects of the town are summed up in the survey in the following terms.

Kiryat Gath has the possibility of developing into a key centre in Israel for textile processing and manufacture, and for food processing; in these industries lies her main future.

This possibility is based on the town's geographical position in relation to the rich agricultural production of the large area surrounding it on the one hand, and to the main markets of the country on the other.

The town has well developed road and rail communications with the rest of the country.

In the setting up of food processing industries nearness to raw materials is of first importance. These industries can work closely with the farm planning authority of the region to ensure a regular supply of good quality raw material.

Kiryat Gath must meet strong competition from other centres, such as Ascalon, Tel Aviv and Rehovot, for the custom of the rural settlers, with the result that the demand of the region for wholesaling, retailing and commercial services is limited.

Much employment could be created if orders for building fittings, etc., needed in the area were placed in the local workshops.

The financial and banking position of the town is much weakened by the fact that firms do their banking through Tel Aviv.

A large amount of industry other than that based on the region because of favourable objective conditions will have to be brought to the town in order to provide employment for its labour force. This will enhance the economic stability of the town and lead to a better occupational structure.

Concentration of industry will draw more industry.

Large-scale unemployment in the town will persist until internal demand and trade opportunities offer a market large enough to take the products of an urban population engaged in food and textile industries.

Closer mutual ties and interests between the town and the region will promote the progress of both.

Recently, Professor Karver, a Director of the Bank of INDIA, who visited Kiryat Gath and the region, observed that the overall plan for its development was properly objective; however, he qualified his approval of the present industry.

In the case of a development town such as Kiryat Gath, Professor

147

Karver stated, which has its demographic, social and economic functions within the surrounding region, the position of labour within such a framework should be clearly defined. He doubted whether at present the labour force in Kiryat Gath is little more than a work colony dependent on the policies, experience and attitudes of the entrepreneur. This in itself is felt to be inadequate in assuring labour stability and, ultimately, urban economic stability. He maintained that industries closely linked with the regional agricultural production should be capitalized in such a way that the primary producer of the raw materials be represented in the enterprise and that a desired relationship be found between the raw material producer, labour and the entrepreneur. It was his impression that where government finance worked hand in hand with private capital without bringing the latter factors—labour and the raw material producer —into overall management, there could arise a situation that financial responsibility towards running the plant would not be entirely met.[1]

Professor Karver's views are especially significant as he has arrived at the same conclusions as those which are the task of our present thesis, though his considerations are of an economic and financial character. We, however, have studied the problem in a broader context, as a result of which we have arrived at the idea of the integration of agriculture and industry, the town and village, harmonizing economic, social and political factors in a single whole.

We are not satisfied with the course of Lachish's development, as it is proceeding neither in keeping with Professor Karver's views nor on the lines we have argued in this study. We shall return to this subject after a description of the changes that have been registered in this region.

The ancient primitive landscape of Lachish has assumed a more civilized mien as a result of unremitting effort in the sphere of land improvement, terracing the hillsides, contour ploughing, afforestation of the water-courses, the planting of orchards and the construction of buildings and roads.

One of the more striking changes that have been introduced is the Zohar Dam, work upon which was taken in hand by the Mekorot Water Corporation in the summer of 1956 and took about a year and a half to complete. The dam, covering an area of 3,000 dunams (750 acres), entirely fenced in, serves as a reservoir for the excess water of the Yarkon River in the winter and spring, and ekes out the Lachish District's supply during the peak consumption period in summer. The water is led via a 48-inch diameter pipe, from the

[1] Quoted from a mimeographed report of Messrs K. G. Abt and M. Black—Farm Planning Department, 1959.

Yarkon-Negev conduit, to a water tower and then via a channel to the Dam. The water flows by gravitation from another dam 3 kilometres (almost 24 miles) to the north. From the Zohar Dam, the water is led—by gravitation—to a station 16 kilometres (10 miles) south, from which it is distributed to other dams in the Negev.

The water serves for both domestic purposes and irrigation. It is chlorinated and purified as it leaves the Dam.

The Dam is owned by the Mekorot Corporation. Bathing in it is dangerous and for that reason forbidden. It is populated with two varieties of fish, carp and tilapia, introduced because they feed on the reeds, which interfere with pumping operations.

The entire face of this region has been lifted and rejuvenated not because of its natural beauty or its historical significance, but in the process of development. That development is based, economically, upon intensive agricultural crops, processed in local factories. Development plans envisaged the establishment of an industrial and services town from the very outset.

Lachish Regional Council was established in 1956 and today embraces twelve *moshavim* and one rural centre—Nehorah.

The council maintains services based upon Nehorah. It has a development committee but no economic institution for the development of the region. Organizationally the council has no contact with Kiryat Gath or any of its enterprises.

The development committee has organizational and representative functions. For example it conducted negotiations with a settlement association regarding the establishment of a cannery. It submitted proposals for the setting up of a filling station and buffet to a local company. It called upon marketing agencies to expand the regional vegetable grading centre. The council decided to erect three concrete structures: two for a shoemaker and barber respectively and a third to be held in reserve. It has joined the Lachish Brook Drainage Authority on which three other neighbouring regions are represented. It has also assisted in the organization of a transport co-operative to which all vehicle owners in the region will belong.

In 1961 the Lachish Region Development Company was established by the council and the settlements. This is a limited liability company with a foundation capital of IL. 2,000 divided into 2,000 shares of IL. 1 each, and reserving 'the right to increase or to decrease the share capital and to issue shares on any part of its local or limited capital, with or without preferential rights, priority or any other right or restriction'.

In addition to the articles usual in such a document the company has inserted two clauses of a specific character in its memorandum of

149

association: 'To acquire by purchase, hire, lease or in any other manner vehicles for the transport of goods or passengers, to receive orders, to hire or to let vehicles of any kind' and 'To establish and to operate the business of tractor station to hire or to let agricultural implements of all kinds'.

Why the development institution should have been set up as a company and not as a co-operative society, only its sponsors know. How it will operate in close proximity to Kiryat Gath, time will tell.

The council embraces only part of the Lachish District. There are at least another two councils in this district, even if we do not take into account parts of others. A glance at Map 1 reveals very strange municipal boundaries. In the centre of the district the borders of three regional councils—Lachish (42), Shafir (40) and Yo'av (36)—come together and separate. The three comprise 45 settlements. Why they cannot form one or, at the most, two regional councils, it is difficult to comprehend. The reason for the existing fragmentation must be sought, it seems, in the heterogeneity of the district. In respect of affiliation, for example: seven settlements belong to the Tnuat Hamoshavim; four to the Ha'oved Hazioni and one to Herut. In the Yoav region are nine *kibbutzim* belonging to two different movements, four *moshavim* and two farming estates. In Shafir there are eleven religious *moshavim*, two belonging to the Tnuat Hamoshavim, three rural centres and one farming estate. In these two regional councils dominant factors are present in Yoav, *kibbutzim*; in Shafir, religious *moshavim*.

If homogeneity is regarded as such an important factor the planners should have paid more attention to the drawing of the regional boundaries in the fifties. But there is not and need not be homogeneity on the basis of movement affiliation. The only region which regards itself as almost homogeneous—Sha'ar Hanegev—and is planning full co-operation, nevertheless has two 'exceptions' among its eleven *kibbutzim*. What harm can there be if the three regions of Lachish are not completely homogeneous? But if three homogeneous blocs—mainly *kibbutzim*, mainly religious, and *moshavim* not affiliated to religious movements—are considered a desideratum, their boundaries can and should be altered. But we have already indicated that we regard the best solution as reorganization on the basis of two regional councils. This is feasible and should be done for a number of reasons. One specific characteristic of this district is that in the areas of the Lachish and Shafir regional councils there are rural centres. In the Yoav region there are none.

But most of the forty-five settlements in these three regional councils are already, or could be, connected with Kiryat Gath. Neither the councils nor the settlements, it should be stated, have ever indicated any concern for the social character of the town. The demarcation of new municipal borders for the three councils, however, is far from essential. In the last resort, of course, such borders are not a decisive factor for what we propose. In more important matters settlements belonging to all three councils—with or without the latter—can join hands. After all, the dividing line between Shafir and the other two regions stems only from the principle of religious education. But in order to set up and maintain a religious school municipal fragmentation is surely unnecessary. In all Israeli towns, including of course Jerusalem, there are both religious and non-religious schools. There is no reason why a religious school should not exist in a united region. There are many spheres of operation besides education.

These three councils, together with six others in the southern district, are at present constructing a joint slaughterhouse, and a freezing and cold storage plant in the Beertuvia region. At one time there was a good opportunity for those selfsame regional councils—three or six or all nine of them—to set up a sugar refinery, before recourse was had to private initiative to establish it. They missed their chance and now they regret it. They could have been the owners of spinning and textile mills, etc. The industry of Kiryat Gath is being established by private initiative, in the heart of a co-operative district settled by *moshavim* and *kibbutzim*. Some opportunity may crop up to remedy the situation, but then it will be far more difficult, and it is doubtful whether the majority of the settlements will be interested in change.

The mistake that has been made will become more perceptible in the near future as the growth of Kiryat Gath changes the face of this agricultural district. It is doubly regrettable that this has happened in a recently developed district, which only a few years ago constituted a *tabula rasa* for planners. We have lost a wonderful opportunity to build an *agrindus* under favourable conditions, because of lack of foresight and vision, lopsided and unintegrated sociological, political and economic planning.

XIII

HEFER VALLEY

W E have chosen to discuss the Hefer Valley region because it presents a number of special characteristics. Among the regional councils it contains the largest number of settlements, it has the largest number of inhabitants and is situated in the geographical centre of the country. It does not have the largest area, however, nor is the council particularly active in the economic sphere, for reasons which we shall discuss here. Our main reason, however, for choosing this region is because of its immense potential for the creation of an *agrindus* in the heart of Israel.

The Sharon includes the entire area south of Samaria down to the northern approaches of Tel Aviv. Sharon may be divided into three sub-districts—the Plain of Caesarea in the north, the Central Sharon or Hefer Valley, and the South Sharon, from Beth Lidd to Tel Aviv. The Hefer Valley has excellent alluvial soil and bountiful sources of subterranean water at no great depth. Dozens of settlements of all types have been established here, the farms prospering on holdings no larger than 20–22 dunams in extent.

Green fodder irrigated with water that is not expensive has enabled the development of thriving dairy-farms. Recently it has been proved that sorghum can be successfully cultivated in this district.

Citrus groves constitute a branch of mixed farming here, while bananas have been introduced, though they are sometimes damaged by frost.

A characteristic feature of this area is the heterogeneity of the soils, resulting in excessive parcellation of the farm-unit—which in any case is not large. The evil effects of this fragmentation are felt particularly in the one-family farms.

The predominant type of soil in the valley is of the brown-red sandy variety, bordering upon sea sand in the west, and heavy clay soils in the north-east. This juxtaposition of different types of soil is

152

typical of all Sharon; it must be recalled, of course, that Hefer Valley is part of North Sharon.[1]

It is a variegated region and its settlements are of all types and belong to all movements in Israel. It has seventeen *moshvei ovdim*, eight other *moshavim* (not affiliated to the Federation of Labour), ten *kibbutzim*, one *moshav shitufi*, six national and regional educational and cultural institutions, four residential quarters—in all forty-six settlements. It covers an area of 110,000 dunams. The soil is fertile, agriculture intensive, and geographically it is very conveniently situated—between the two major urban centres of the country, Tel Aviv to the south and Haifa to the north. It is served by good roads and its municipal borders are very favourable. It is interesting to note on Map 1 how tortuous are the borders of regional councils

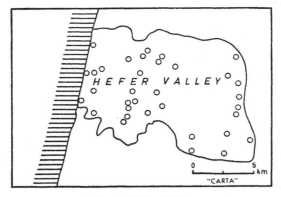

Map 8. Hefer Valley

to the north and south of Hefer Valley, whereas the Valley itself is an ideally contiguous area (see Map 8).

Hefer Valley indeed was one of the four regional councils set up by the Mandatory government—as far back as 1940—as even in that period the need for some municipal organization of the area was apparent. After the establishment of the State of Israel the borders of the region were altered several times and are today as follows: Natanya to the south; Hadera to the north; the sea to the west and the Jordan border to the east. The settlements established for security reasons have not been included in its jurisdiction but will be, certainly, in the near future.

Since the foundation of the State twelve new settlements have been set up, occupying a total area of almost 30,000 dunams, while 21,000

[1] H. Halperin, *Changing Patterns in Israel Agriculture*, Routledge & Kegan Paul, London, 1957, p. 18.

dunams have been allocated to previously existing settlements to enlarge the land holding which previously stood at 20 dunams to 27 dunams in the *moshavim* and 25 dunams in the *kibbutzim*.

Changes had also to be introduced into the water economy. Hefer Valley is bountifully endowed with water resources, but decades of intensive irrigated farming and excessive exploitation of wells have led to a steady lowering of the water table and increasing salinity. The solution proposed presupposes regional organization (for which reason we mention the problem in the present context) to be integrated in a national water scheme with a view to rationalizing dwindling local sources. The Alexander Stream, rising in the mountains near Nablus, and fed by smaller tributaries, flows through the district. In the winter months the brooks burst their banks and inundate neighbouring fields. Thus a comprehensive drainage scheme is necessary. When this area was first settled in the thirties the settlement authorities undertook this task. Today it has reverted to the regional council, which maintains a grader for cleaning internal channels. A branch of the field service has also been set up to ensure effective use of water.

The settlements handle the sewerage problems individually. The council, however, is responsible for the maintenance of approach roads and for the construction of new ones. Its budget for this purpose being very restricted it seeks Government grants, its own contribution being hardly more than nominal. The council also engages in improvement of the sea front. It obtains loans upon its own responsibility for the erection of school buildings. It has constructed an amphitheatre and a museum, and maintains avenues of trees along the approach roads. Children's nutrition schemes, the youth hostel, and similar activities also come within its scope.

There is of course no doubt about the value of the municipal activities and the services instituted, but the Hefer Valley Council, the largest regional council in the country, has not initiated a single enterprise for processing, or for that matter, for storage or curing, of farm produce. It is hardly necessary to add that it has not planned any industrial undertakings.

It may be of interest to note here that while the manuscript of this book was undergoing its final revision the Council was contemplating the establishment of a margarine factory, involving a large investment, using imported raw materials and capable of providing employment for only a small number of workers. The writer strongly advised against taking the project any further and succeeded in persuading the Council that a services enterprise or a plant processing locally grown raw materials would be far more in consonance with the Valley's needs.

There are a number of enterprises in the various settlements. One *kibbutz* operates a bakery, a second a garage, a third a cannery, a fourth a plastics factory, a fifth a barrel plant, while there is a printing works in one of the *moshavim*. None of these is based upon any inter-settlement collaboration; in none has the region any direct interest. The reasons are (*a*) the ten *kibbutzim* in the area are members of the Hefer Valley and Samaria Settlements organization, already discussed in Chapter VIII; (*b*) the *moshavim* have not shown any inclination to join this organization, or any initiative to establish partnerships between themselves or within the framework of the region.

This, of course, does not mean to say that regional co-operation is a lost cause in the Hefer Valley.

Objectively there are very good conditions for the establishment of an *agrindus* in this district. Indeed, if and when an *agrindus* movement is set in motion in Israel the Hefer Valley may be one of the foremost among those putting its principles into practice.

XIV

THE B'SOR DISTRICT

THE western section of the Negev is traversed by the B'sor Stream, the main tributaries of which are Wadi Sabe, Wadi Sharia, Wadi Nakhbir and Wadi Assaf. The B'sor District is not a definite geographical entity and indeed it has only been named, after the stream passing through it, for the purposes of planning new settlement (see Map 9). It covers an area of 733,000 dunams (183,250 acres).

In the municipal map of Israel (which we have used so extensively in the course of the present study), the B'sor District comprises—and goes beyond, to a minor extent—the areas of the Maon, Azata and Merhavim regional councils. There are forty-one settlements with over 10,000 inhabitants in the three regions. It has been estimated that 683,000 dunams of land in the B'sor District are suitable for cultivation. 240,000 dunams of this area have already been allocated to the settlements in the region. New villages may be founded on the remaining 443,000 dunams.

The climate is dry, average annual rainfall dropping from 360 millimetres in the north to 130 millimetres. In the hottest month of the year the average temperature is 34° C. in the eastern sector and 31° C. in the west.

Climatogically B'sor can be divided into three sub-districts. The whole district has a high solar radiation, an important factor in stimulating the growth of various crops in the winter, though in one of the sub-districts, the Beersheba highlands, there is the danger of frost in the winter. Otherwise the entire area is excellent for early ripening varieties of fruit and vegetables.

The cultivable soils are of the loess, sandy, or sand on loess types. From the physiographic and geomorphological points of view the soils of this District can be classified in seven categories, subdivided into twenty 'soil groups'.

Water. The expansion of agricultural settlement, and the development of the District generally, will be feasible only if water is made available through the main conduit leading from the Jordan, and the completion of all water projects in Israel, including floodwater conservation, and sewerage schemes. The Jordan–Negev Project allocates, in the final stage, 130 million cubic metres to the B'sor District.[1] Most of the water will be piped over a long distance from the

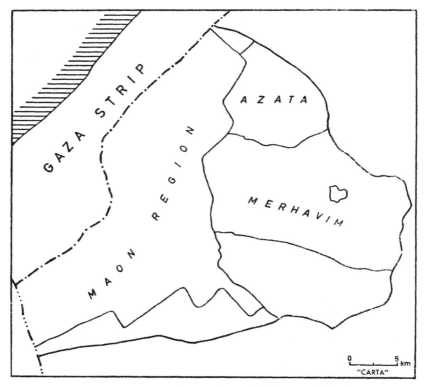

Map 9. B'sor Region

north; only a small quantity will come from damming local floodwaters. The flow of water in the watercourses is of a typical desert character and heavy and expensive equipment will be necessary for long-term conservation. This equipment will catch and retard the flow in the watercourses, lead it to an aquifer and purify the water before it enters the main pipes. Some of the water will be used

[1] From an Israel Water Planning Authority Report, submitted to the Minister of Finance and Head of the Settlement Department.

immediately for flood irrigation of the fields. Dams, reservoirs and conduits must be constructed.

The investment involved is very considerable and for this and other reasons regional planning is essential. It has been proposed that the methods applied in the Lachish District be used in B'sor. Rural centres will be set up to serve clusters of 5–6 villages in which there will be 350-450 families. Concentration of the services for a number of villages and hundreds of settlers will, of course, lower their cost. The rural centres will be the location of marketing and grading, health, educational and cultural institutions, the tractor station, etc.

There are already approximately 2,000 farm units in B'sor. The holdings in existing settlements will be increased by 800, while another 4,500 will be established in the new villages projected. There will be about 2,000 families in the six rural centres. The two existing towns of Ofakim and Netivot with a combined population of 1,750 families will be converted into rural centres, and another 3,000 families will be settled in them. A new town of B'sor will be the capital of the B'sor District. It has been estimated that the district will have a population of 24,000 families or about 100,000 inhabitants.

The project for settling the B'sor District was first brought to public notice by the Head of the Jewish Agency's Settlement Department (who is also the Israel Minister of Finance), who issued the slogan 'The B'sor District is the challenge of our generation'. In the course of a comprehensive and exciting address, he spoke of the plans for the town of the District—in which we are specifically interested in this work—in the following terms:

'In the centre of the B'sor District will be the provincial town— B'sor. The town will ensure the repopulation of the Western Negev, which today is still empty. It will constitute the centre for service and other installations and for the industrial enterprises required for successive stages of processing the agricultural produce of the district, such as canneries, packing houses and cotton gins. Within the planning of B'sor the existing development towns of Ofakim and Netivot (Azata) will be included. This, there is no doubt, will give a new stimulus to their progress.

An important factor in the siting of the town of B'sor is that it is located on the east–west axis of the Negev, which runs from the port of Ashdod, via Sha'ar Hanegev, to Gevulot and Revivim, thence to Sde Boker, cutting the road to Eilat.

The contour plan is for a town of 50,000 inhabitants in the final stage. For the time being 10,000 people have been planned for.

In the initial phase a textile combine will be established in B'sor. This combine, to comprise finishing, dyeing and printing plants, will

158

supplement the spinning mills in the Negev and Lachish and the weaving mills in Kiryat Gath, Beersheba and Ofakim.

In the course of time the necessary specialists will be found to raise textile printing to the highest standards of fashion. In B'sor there will also be factories for citrus preserves and concentrates, and for the canning of fruit and vegetables grown by the Negev settlements.

Plans for other industries, to be established and operated both by private investors and by the settlements (through the agency of development companies set up by the regional council, or in some other fashion), are at present being studied. In this manner the settlements will contribute towards the development of the town. Thus the ties between them will be based not only upon the services available in the town but upon the processing of their products. Such ties can prove extremely advantageous to the townsmen no less than the farmers. The economic and social principles applied in agricultural settlements are suitable equally for the town. We venture to hope that relationships based upon co-operation and mutual aid will be introduced into the town, just as they are already the rule in the village, and that they will contribute towards its economic and social prosperity.'

In the course of his remarks the Minister dealt with the security aspects of the settlement of this desolate region in the southern part of the country, facing the Gaza Strip.

Listening to this address I thought that little was lacking to attain the goal of *agrindus*. That little must be made up no less in Upper Galilee, in the Jordan Valley, the Hefer Valley and Sha'ar Hanegev. It is the fate of any important idea that in the process of implementation it is not improved, but complicated by the conditions under which it is put into effect and the obstacles that must be surmounted.

The Minister's address was suffused with a spirit of co-operation. But though he declared that 'the economic and social principles applied in agricultural settlements are suitable equally for the town', he weakened the impression of his introductory remarks, making the central idea, which is of interest to us, vague. The team of planners engaged on the detailed programme[1] for the settlement of the B'sor District have devoted a special section to the town of B'sor. Here the outline given in the Minister's speech is repeated, though in greater detail, particularly in regard to geographical location, roads, area, the number of spindles in the mills, the quantity of fruit to be processed in the canneries, and the number of workers

[1] The B'sor District, general physical data. Outline for Settlement Planning. Jewish Agency: the Settlement Department, Jerusalem, July, 1961.

who will be required in each phase of development. The body of the project is carefully planned and figures abound, but the soul is lacking—the social idea, the social organization of the town, and the social character of the population. There is the risk that this town will be constructed upon the lines of Kiryat Gath. But this is a prospect that can be averted, though the task will not be easy.

In the B'sor District, we have pointed out, in addition to agricultural settlements, there are two development towns, Ofakim and Netivot (Azata).

In the Lachish District there are three types of settlements, representing successive levels—the village, the rural centre and the town. In the B'sor District there will be four—the villages, rural centres, two townships and a town.

We have already referred to the fact that the planners propose to integrate the two townships and to develop them as service centres for the neighbouring villages, but it is not clear what form they are to assume. The two towns today have 8,000 inhabitants and are planned for 21,000. The plans state that grading and packing sheds for farm produce are to be put up in both, and immediately afterwards follows a passage saying that 'agricultural and industrial development will, it is estimated, enable the two towns to absorb another 13,000 inhabitants . . .'

It is obvious, accordingly, that they will not be merely rural centres, but townships. So there will be two townships and one town in the district. Why?

Let us see what the plan has to suggest in regard to the rural centres:

'The six rural centres which have been planned—either as entities separate from the agricultural community or as part of a large composite village—are designed to serve and to meet the needs of farmers. In these centres workshops and even small industries will be established. But their principal function will be to supply services of an adequate standard to the settlers. In these centres sheds for grading and packing vegetables, packing houses for citrus, dry and cold stores, garages, metal workshops and tractor stations, as well as educational, cultural and medical services, will be concentrated.

Each rural centre will count 300–320 units (families) and there will be 2,000 units in all in the rural centres. Some of these families will support themselves by providing services for the settlers, others will work in the agava plantations, in the citrus groves and in the workshop which will be set up in the rural centres.'

That is all there is in the plan.

The planners, who, it seems, are engaged exclusively in preparing

160

the physical blueprints, have little interest in the relations that must develop between the inhabitants of the rural centres and the villagers. This is the concern of other planners. It is interesting to note that the Housing Authority does not disregard the social problems, and employs sociologists, research workers and instructors. The settlement authorities, too, will have to operate in these fields at some time in the future. But for some reason, the preliminary planning has not taken cognizance of these aspects of settlement in the district.

The rural centres will be populated by 2,000 families. In the two existing townships there are today less than 1,500. Why should they be expanded threefold? It seems to us that it would be more practical to convert them into two of the six projected rural centres. The present inhabitants of these two townships would be sufficient to establish another two or three, which today exist only on paper. Thus Ofakim and Netivot will contract and not expand. And instead the town of B'sor will develop to its planned proportions within a shorter period than foreseen. In this town administrative, organizing, educating and guiding forces will be concentrated, instead of being spread over three towns, all of which are liable to be problematical. There need be only one town—B'sor.

Internal 'migration' within the area will prove less difficult than bringing new settlers from elsewhere. Only one-sixth of the inhabitans of Ofakim have been settled there for five, and one-third for no more than two, years. We shall not be uprooting and transplanting established residents. Indeed, the operation will be of a very mild character. Sections of the population like these often pull their own roots up and move off to some new, more distant ground. Let this section move to the town of B'sor, if it wishes to do so.

The economic significance of the settlement of the B'sor District is very interesting. The plans provide for the development of a special type of agriculture, producing mainly for export—citrus, winter potatoes, early ripening melons, ground nuts—and for replacing imports, such as sugar-beet. Other crops will be intended for the local market: early vegetables, winter tomatoes and peas for canning. About 80 per cent of the estimated production—valued at about IL. 76 million—will be exported or will replace imports.

The investment capital required for the agricultural settlement of the B'sor District has been estimated at IL. 150 million. Two-thirds of this sum will be required for means of production and the balance for housing, public buildings, etc. This sum does not cover the costs of water development. Of the 700 million cubic metres which the national water plan is expected to add to existing resources—at a cost of IL. 550 million—130 million have been earmarked for the

B'sor District. This gives us some idea of the investment required to bring water to the B'sor District.

In addition to the money to be sunk in agriculture, funds are required for the development of industry and services. The investment per worker in industry has been set at IL. 16,000 and at IL. 10,000 in services.

This settlement project will provide employment for 20,000 earners. The ratio between those permanently employed in agriculture and those working in industry, services and seasonal farming jobs will be 1 : 3; in other words there will be 5,000 agricultural units and 15,000 earners engaged in other occupations.

In the settled areas of B'sor certain crops have been successfully cultivated. The lemons are the best in the country. The grapefruit is good, and at least equal to that grown elsewhere in Israel. About the oranges we shall only be able to speak in a few years, as the groves are still young.

The ground-nuts are excellent, yields being higher than elsewhere. This is also the case with winter potatoes and tomatoes, early onions, cotton, sugar-beet, carrots and peas. It is hoped that the southern sector of the District will not lag behind the northern once water comes.

Investment—both for production and consumption—in a farm unit has been put, in the plan, at IL. 38,000. Of this sum the settlement authorities will provide 80 per cent and the balance will be paid by the settler, out of his own capital or savings. Such a farm should bring in a net income of IL. 6,200–6,800, out of which the settler must pay off his debts to the settlement authorities. The income that remains will not be large but it will be sufficient.

In the budget of the unit, however, one important item is wanting, and it is essential that adequate provision be made for it in a district in which it is planned that agriculture, services and industry are to be integrated. For practical and social reasons it is essential that the settler should be a partner in regional enterprises, whether of a service or industrial character. To enable him to participate in the rural centre and in the town, his allocation must include a sum for the purchase of a share, whatever its price. But what is even more important is the social significance of the settler's participation in regional economic company or society, symbolizing the integration of agriculture and industry in the region.

There are, accordingly, three changes that must be made in the plans.

(1) There must be only one town, B'sor. The townships must be absorbed in the plans for the rural centres and for the town.

(2) The town must be of a co-operative character. Its enterprises

must be owned co-operatively by the agricultural settlements of the region, with the participation of societies of permanent workers employed in the town's enterprises, who are not members of the villages.

(3) The agricultural settlement grant must include a sum to enable the settler to purchase a share in the economic company or society for the development of the region—the future B'sor District Development Society—which will establish and manage the industrial and service enterprises in the region.

XV

OTHER REGIONAL COUNCILS

LEADERS of the regional councils constantly complain about the lack of suitably qualified candidates to launch and manage enterprises. It is of course essential that they be members of settlements in the district, but the settlements need them themselves and are reluctant to permit them to engage in work on behalf of the region as a whole. There are no easy solutions to this problem. Leaders, managers, must grow with the projects they head. This is a process that takes a long time. If, however, our proposal for the re-drawing of the municipal map were accepted and the fifty regional councils reduced by two-thirds, at least one hundred local leaders would be freed for new tasks (for each of the thirty-three councils is headed by a chairman and two deputies). This is an important human asset; it could be utilized to set up and manage at least a hundred large, new undertakings. This number could be augmented by the amalgamation of branches and services.

But difficulties of a personal and psychological character make such suggestions appear wishful thinking.

We have discussed one large regional council at some length. We have dealt more briefly with another four. It is, of course, superfluous to outline the others, however sketchily.

However, seeing that in the last few chapters we have concentrated upon the southern district, let us go north and cast our eye over five regional councils—comprising from five to eight settlements each—which should be united to form a single council. We are referring to the Hevel Yavneh, Gederot, Nahal Sorek, Brenner and Gan Raveh councils, which together cover thirty-three settlements (see Map 10). Such a union would liberate much creative, constructive and managerial ability, which could fruitfully be invested for the advancement of this large—but not overlarge—region.

Going eastwards we come upon two other regions—Gezer and

Map 10. A projected *agrindus* in the south

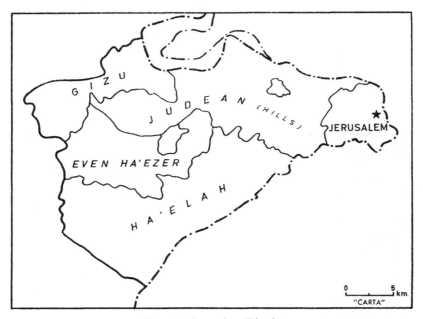

Map 11. Jerusalem District

Modi'in—from which a successful *agrindus* counting forty-three settlements could be constructed.

And even in the Jerusalem District (Jerusalem is a city set apart) the agricultural area surrounding Jerusalem is broken up into four regions—the Judean Mountains with twenty-five settlements; Ha'elah with thirteen; Even Ha'ezer with ten and Gizu with nine. We have suggested unification of these four regions to local leaders. We asked them why Beit Shemesh, which is almost in the exact centre of the combined area, should not be the regional town. The proposal, we were told, was a reasonable one, but it would be better to build a regional centre elsewhere, nearer Jerusalem, as the civic leaders of Beit Shemesh would not agree to join such a union. Our

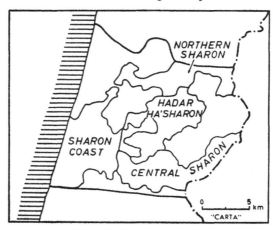

Map 12. Sharon *agrindus*

statement that Beit Shemesh is situated virtually at the geometric centre was countered by psychological and practical arguments, fortified by the contention that seeing that in any case Jerusalem was the main market for their produce, the centre for processing and storing it should be as close to the city as possible. This applied equally to non-agricultural enterprise, they said. These are considerations that must be studied. But what is most important is the prospect of combining four adjacent regions and perhaps of forming them into an *agrindus* (see Map 11).

Further north, we have seven councils with winding borders and comprising from four to twelve settlements. It is difficult to offer a suitable solution for the three councils in the southern sector, but why should the four Sharon regions—Hadar Hasharon, Sharon Coast, Central Sharon and Northern Sharon—not be united? The

166

combined area would have forty-one settlements, and even if the existing local councils should continue to constitute enclaves, one large regional council, and even an *agrindus*, could be formed (see Map 12).

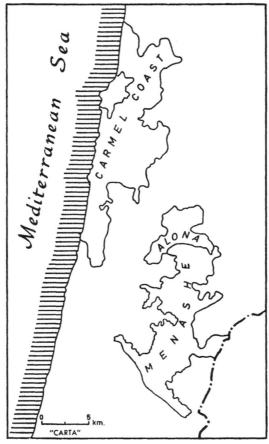

Map 13. Çarmel Region

North of the Sharon comes the Hefer Valley which we have already discussed in a previous chapter. Further north boundaries form a maze, mainly because of the enclaves of local councils. Near Hadera and Gan Shmuel is the 'industrial zone' of the 'Hefer Valley and Samaria Settlements', with which we have already dealt elsewhere.

Then come two adjacent councils (see Map 13). That of Alona,

A.—M   167

which has only three settlements, should be done away with by bringing them into the Menashe council, with its eighteen settlements. If it were not for the large number of local councils and the inter-regional (from Tel Aviv to Haifa!) industrial zone, this region with its twenty-one settlements would be eminently suitable for a medium-sized *agrindus*. We ourselves, however, would prefer to leave this region for a later stage. Perhaps, at some time in the future, when the country has a more rational development policy and, more important, if and when there is a more favourable climate for the *agrindus* idea, we shall be able to consider this awkward area too.

In the neighbouring region of Carmel Coast the situation is hardly any better. Here, however, the difficulty derives from topographical conditions, from the hills and valleys which break up the district. Some of the *kibbutzim* in this area are members of the Hefer Valley and Samaria Settlements organization. But the important activities which have been embarked upon in this region including drainage, a field service for rational use of water, water supply, a tractor station, maintenance of internal and approach roads, two regional schools, and the construction of an amphitheatre, etc., already constitute important factors for replanning in the future. The establishment of an *agrindus* in this region counting twenty-two settlements is certainly not out of the question.

Going northwards, beyond Haifa, we come to the regional councils of Zevulun with ten settlements, Na'aman with five, Ga'aton with fifteen and Sulam Zor with twelve—four councils with a total of forty-two settlements (see Map 14). The long strip of coastland they cover, it is true, can hardly conveniently be united in a single framework, but it could easily be fitted into the jurisdiction of two. Again there does not seem to be any reason why two *agrindi* should not develop here.

Eastwards we have the Megiddo region with twelve settlements, Kishon with fifteen and Jezreel with eleven (see Map 15). It might not be feasible to unite all three, but it seems possible that the number of regional councils could be reduced to two. Social conditions in this area are certainly suitable for the setting up of two *agrindi*.

We have not considered the position of the Lydda Plain region with ten settlements in the heartland of Israel.

The future development of the Ma'aleh Hagalil region (in Galilee) comprising no more than six settlements should be studied and the possibility of a merger with Merom Hagalil with fourteen should be considered.

There is also the Bnei Shimon region in the Negev, which includes eleven settlements. The regional council has floated a 'Bnei Shimon Regional Development Co.', which has already invested a total of

168

IL. 300,000 in agricultural machinery, etc.—cotton pickers, shovel-dozers, equipment for spreading fertilizers and for cultivating sugar-beet, a weigh-bridge, etc. When this area is more closely settled an independent *agrindus* might be established.

From the social point of view the Ma'on region resembles Sha'ar

Map 14. Galilee Coast

Hanegev closely and there is no reason why they should not be merged. Today, for planning purposes, the former comes within the B'sor District, but where it should ultimately be—within B'sor or united with Sha'ar Hanegev—should be reconsidered at some future juncture. It cannot continue as a separate entity for long. Up to the present, it is true, it has developed certain agricultural services,

169

heavy equipment and machinery, a cold storage plant and a grading house, but the council is in difficult financial straits, because it is not capable of carrying even the present restricted number of enterprises.

Finally we have still to consider two regional councils south of the Jordan Valley—Beit She'an and Gilboa (see Map 16). The Gilboa region occupies a special place in the development of the *kibbutz* movement. Its resettlement began in 1921 and it was here that the

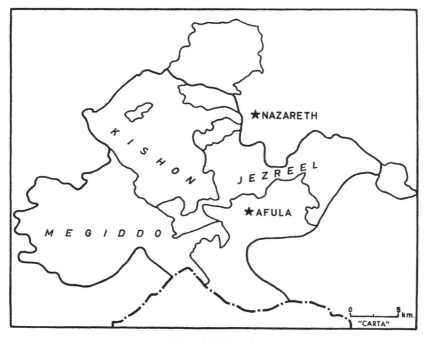

Map 15. Jezreel

first large collective village—Ein Harod—was founded. Gilboa, indeed, was the cradle of the G'dud Ha'avoda, the location of the first *kibbutz* of the Hashomer Hazair movement—Beit Alfa, and of the second *moshav*—Kfar Yehezkel. At one time the centres of two major collective movements—Hakibbutz Hameuchad and Hakibbutz Ha'artzi—Hashomer Hazair—were situated here. Here, too, the integration of agriculture and industry was first put into practice. Gilboa was one of the first four regional councils to be established. Today it comprises twenty-two settlements, with a total population of 10,000 inhabitants. Indeed, in point of number of settlements and inhabitants, it is the third largest region in the country, surpassed only by Hefer Valley and Upper Galilee.

Approximately 140,000 dunams of a total area of 240,000 dunams (60,000 acres) are under cultivation.

The regional enterprises are a grapefruit packing house, co-operatively owned and operated by the growers in the district, and an association of eleven settlements for the operation of heavy equipment. Two other enterprises are at present under construction: a

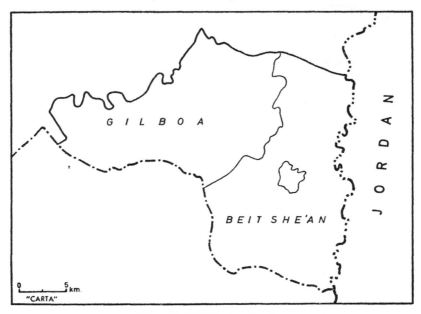

Map 16. Beit She'an Region

slaughterhouse in which all settlements of Gilboa and neighbouring Jezreel Valley are partners and a co-operative garage.

The regional council retains an interest in the co-operative societies through the stock it holds.

The rules upon which the local slaughterhouse has been founded resemble those subscribed to in other regions:

(*a*) The slaughterhouse society is called 'Of-Kor'.

(*b*) The society conducts its financial operations upon its own responsibility. The settlements guarantee the operations of the society.

(*c*) Every settlement invests an equal sum as membership dues in the society. This sum is IL. 5,000.

171

(*d*) The society undertakes to accept all table poultry supplied by the settlements.

(*e*) The settlements undertake to offer all their table poultry exclusively to the society. The society will send live poultry to the market as it sees fit.

(*f*) Profits made or losses incurred by the despatch of live poultry to the market are put to the society's account.

(*g*) The directors of the enterprise decide upon the grading of birds and the minimum quality accepted for processing.

These rules have been generally accepted as a sort of model by regional societies of this kind.

This region was one of the first in Israel to establish a water society and a regional drainage authority. There is also an association for the preservation of wild life, which has demarcated a nature reserve on Mount Gilboa and repairs the damage done by roaming herds of deer. North of Gilboa is the Beit She'an Valley region, which borders again with the Jordan Valley.

No region is like its neighbour or, indeed, like any other in its structure or the development of joint enterprises. A special feature of the Beit She'an region is two co-operative undertakings of an agricultural character, in addition to joint enterprises of the type we have seen elsewhere. These are a citrus grove and fish-breeding pools.

The Beit She'an Valley region comprises 152,000 dunams of farmland, upon which are thirteen *kibbutzim*, five *moshavim* and two farming estates of the Yitzur U'fituach Co. owned by the Ihud Hakvutzot Vehakibbutzim.

The water economy of Beit She'an, too, is of a special nature. Of the 70 million cubic metres available 50 per cent are too saline for normal agricultural purposes and are used for fish-breeding ponds. The latter cover an area of 37,500 dunams. Eighteen thousand dunams are given an auxiliary irrigation, while 74,000 are under un-irrigated cultivation.

The Beit She'an Valley is a level plain below sea-level. Temperatures in the valley are generally high, though frost, causing much damage in banana plantations and tomato fields, is not unusual during the winter nights. During the summer evaporation is high, preventing the use of ordinary sprinklers, as the fine spray emitted does not saturate the highly permeable chalk soil, which requires large quantities of water. Most of the area consists of calcareous soils, the chalk content being higher here than elsewhere and exceeding 70 per cent. A large part of these calcareous soils are saline—sometimes to a degree preventing cultivation. There is also brown soil in the district which is more suitable for cultivation.

Water for irrigation is supplied by the numerous springs at the foot of Mount Gilboa. The water of the larger springs, however, is saline and contains a high percentage of calcium carbonate ($CaCO_3$). But in addition to this, we encounter in this region the phenomenon of salinity of large areas which would normally cause serious agricultural problems.

These problems were solved by the establishment of carp-breeding ponds, this fish developing satisfactorily in saline water and enjoying especially the warm climate of the district.

The establishment of fish-breeding ponds in this region is fully justified even from the agricultural point of view, although limits should be set to an expansion of this branch of farming which requires large areas of land and a considerable amount of water.

In the light of these considerations carp-breeding will have to reckon with legitimate opposition from the Ministry of Agriculture, especially if the farmers of the district should emphasize the trend to extend fish ponds over areas of good land and sweet water. Such a development indeed may react unfavourably upon the density of population in this border zone, which has favourable agricultural prospects.

Here, too, as in the neighbouring Jordan Valley, lucerne is the major irrigated crop. In spite of this the expansion of dairy-farming has been retarded because so far no breed of cattle has been satisfactorily acclimatized. The lucerne, accordingly, which flourishes in this district, is ground into meal, finding a ready market among cattle and poultry breeders in all parts of the country.

Other field crops, including winter grains, are all in need of supplementary irrigation to secure a yield equivalent to that obtained elsewhere on unirrigated soil. The major factor responsible is the inefficient and primitive methods of irrigation which cause salination of the soil.

Vegetables can be raised successfully on suitable land with proper methods of irrigation, rational crop rotation and the eradication of pests and diseases. Potatoes grown here enjoy the important advantage of being the first in the market.

In regard to fruit-growing, this region is not among the most suitable; for example the date and the pomegranate grow better in other fruit-growing districts. Again both citrus and vines suffer from salination of the soil, while the banana, which flourishes near by in neighbouring Jordan Valley, has to contend here with two enemies—saline water and frost.[1]

[1] H. Halperin, *Changing Patterns in Israel Agriculture*, Routledge & Kegan Paul, London, 1957, pp. 12–13.

In recent years cotton growing has been undertaken on a considerable scale in the valley. The results have been highly encouraging.

Because of the semi-arid climatic conditions of the Beit She'an Valley, priority was given to a regional water scheme. The latter served as the model upon which other joint enterprises were undertaken.

The water project, it is true, was planned by the Israel Water Planning Authority (Tahal) and executed by Mekorot, but the regional council was in effect the motive force behind it, persuading all the local settlements to accept the joint plan. The council regards the undertaking as its own and supervises the allocation of water.

There is also a co-operative society for agricultural machinery and implements which has purchased large tractors for earth-moving, a shovel-dozer for loading manure, a lucerne planter and a large spray for orchards. The society also maintains a regional station for spraying cotton and other crops. There is a regional garage to service this equipment, as well as the machines and trucks owned by the settlements.

The first lucerne meal mill was established in the Beit She'an region; so was the first jointly-owned (by the regional council and the Yitzur U'fituah Co.) cotton gin. In 1960/61, 6,350 tons of raw cotton and 2,307 of fibre were ginned here.

The region also has a date curing and packing house, so designed that it can be enlarged in phases to cope with the growing crop. Another undertaking is the experiment station maintained by the regional council to test the suitability and prospects of acclimatization of various crops.

As indicated the Beit She'an Valley is distinguished from all other regions in Israel by the phased development of two agricultural branches as joint ventures.

Few of the local settlements possess land suitable for the planting of citrus groves. Citrus growing, however, could make a substantial contribution towards maintaining the agricultural stability of the settlements both as a source of extra income and the employment it would create. Areas suitable for groves, however, are situated at some distance from the settlements, too far, indeed, for them to contemplate engaging in citriculture individually. For these reasons the regional council resolved to plant a grove, jointly owned by all the settlements in the district.

The State Lands Commission allocated an area of about 3,000 dunams for this purpose, on the basis of two dunams per farm unit.

The other joint projects of this character are the large fish-breeding ponds, covering almost 3,500 dunams.

It seems to me that the two excellent regions—Gilboa and Beit

She'an—with their forty settlements could successfully co-exist in one united *agrindus* with the town Beit She'an as its city.

HIGHLAND SETTLEMENTS

From the agricultural point of view highlands are submarginal areas. The reason, of course, is not because of their inconvenient topography. The highland districts in Israel indeed possess a number of advantages, for example, climatic conditions that are good for man and beast, and also for fruit-growing, and extensive grazing lands. But these are more than outweighed by the disadvantages. The soil, in the majority of areas, has little depth and yields are low. In addition, most of the soils are calcareous, agricultural use of which is restricted. The soils, moreover, vary sharply in character, the areas are not contiguous and the plots are small and fragmented both in point of altitude and distance. Cultivation of the slopes is difficult and land improvement is a never-ending task. Stones must be removed from the land and many ploughings are necessary. Generally speaking a far greater investment, in both money and labour, is necessary for the development of orchards in the highlands. Because of fragmentation of the cultivable land and the high proportion of land that is not suitable for tillage, the agricultural unit must necessarily be large, which puts obstacles in the way of bringing the farms together to form a village unit. Transport is expensive and the extra expense of pumping water from the valleys to the hills makes the costs of irrigation almost prohibitive.

Objective circumstances, however, compel Israel to settle the hills despite these drawbacks. If it is true that of the continental area of the globe the flora-covered flatlands constitute no more than 20 per cent, with 32 per cent given over to deserts, 30 per cent to forests and 18 per cent mountains, the highland area of Israel is almost equal to the world average. The hill country comprises 4·5 million dunams (about 1·1 million acres) of a total area of 21 million dunams. Israel, it will be seen, is not a mountainous country, but mountains take up a considerable part of its area. The highlands require much care and devotion. It took two generations of unremitting work before the 'easier' areas—agriculturally speaking—were improved; before the swamps were drained, fever eradicated, irrigation installed and deep-rooted weeds brought under control in the valleys, in the plains and along the shores of the Mediterranean. But the highlands, too, were not forgotten, though to this day no feasible method of highland settlement has been put forward. (It is doubtful whether a suitable solution has been found in any country of the world.) In spite of this, however, we have today 81 well-organized highland villages—59

175

moshavim and 22 *kibbutzim*—with a population of 4,200 families (20,000 inhabitants) out of the total number of 7,200 planned for. They engage mainly in fruit-growing, their orchards covering two-thirds of the cultivable land. The superiority of highland fruit is emphasized; the highland apple contains 25 per cent of dry matter as against 14 per cent in that of the valleys and the sugar content of the former is twice that of the latter. Highland fruit has a better flavour and colour, and is more suitable for storage. The only drawback is the high cost of water.

Poultry-runs in the highlands are stocked with 1·2 million layers—which, the planners insist, should be brought up to two million.

These are the twin pillars of highland farming. Some experts are sanguine about late autumn vegetables. Other branches engaged in, though on a more restricted scale, are tobacco, field crops, sheep- and cattle-breeding and bee-keeping.

The net income of a highland farm is about half that of a farm in the lowlands.

As long as agricultural technology has not discovered how to establish profitable highland farming, the corrective, we believe, may be sought in the development of a number of State experimental farms, in the establishment of vacation centres with convalescent and rest homes, boarding houses and hotels, in all of which the sons and daughters of the settlers can find employment. Above all, however, we think that regional industrial enterprise holds out the best prospect for settlement of the highlands. Smallholdings, confined to a few branches of farming, can eke out the wages of the factory worker.

Most of the highland settlements are concentrated in the Jerusalem Mountains and are organized within four regional councils. None of the latter has so far set up an industrial undertaking. They will not do so themselves, for the villages are weak and poor. This is a task for the settlement authorities; they must promote integrated agricultural-industrial settlements.

We need have less anxiety about the eight *moshavim* and twelve *kibbutzim* in the highlands of Upper Galilee. We have already discussed the important work done in this region (in Chapter IX). But the situation of the settlements of Central, Western and Lower Galilee is very similar to that of the Jerusalem villages.

XVI

DEVELOPMENT TOWNS

THE headlong flight of the Arabs from Palestine which was one of the early consequences of the outbreak of the Arab-Jewish War, and the unprecedented mass-influx immediately after the establishment of the State of Jewish survivors of the European concentration camps and refugees from the Islamic countries of Africa and Asia, generated a demographic revolution in Israel. The effects of this revolution are still apparent in every sphere and every walk of life in this country fourteen years after the foundation of the State, and will continue to be so at least during the lifetime of the present generation. The arrival of hundreds of thousands of newcomers, the majority of whom were completely destitute, called for the adoption of a number of immediate measures. New conditions gave birth to new types of settlement. The *maabara* (transitional work camp), the *moshav olim* (immigrants' settlement), the *kfar avoda* (labour village) and the development township emerged.

The basic principle adopted in handling the influx can be summed up as 'From the ship (or the aeroplane) to work'.

The camps (*maabarot*) were planned primarily with a view to enabling the new immigrants to work for their living, while at the same time they facilitated the development of somewhat better housing conditions—large tents in the first phase; aluminium and canvas huts later—the construction of paved paths between the tents or huts, and even the cultivation of small plots of vegetables. The residents of these camps were employed in the vicinity and constituted a reserve of candidates for the establishment of permanent settlements—of either a rural or urban character.

From the *maabarot* one stream was led to agriculture by way of the *kfar avoda* and the *moshav olim*. Later a system of bypassing the *maabarot* and transferring the immigrants direct from 'Ship to Village' as well as to the development townships—which we shall discuss in the present chapter—was evolved.

177

There is no question that, in view of the immense proportions of the influx, every prospect of housing, employing and supporting the immigrants had to be exploited. It was regarded of paramount importance to make the utmost use of any niche in the country's economy. There was no time to plan and to test plans in the light of later experience. Entire towns had been abandoned by their former residents. It is true they had suffered severely in the course of, and as a result of, that flight, but there remained buildings, dwellings, which could be put to use. At the same time the settlement authorities called upon the workers' settlements to forgo, temporarily at least, their principle of self-labour in favour of the equally sacred rule 'that thy brother may live with thee'. The settlement movements decided to accede, and the established villages found *maabarot* spring up overnight (literally!) in their neighbourhood, and hundreds, sometimes thousands, of newcomers clamouring for work. Gradually the residents of these work camps were transferred to housing estates, which formed the nucleus of new towns. Many of them settled down permanently in the abandoned towns, which were now being repopulated. The absorption authorities proclaimed a number of old towns with a sparse population as 'development towns', granting important facilities in the fields of housing, employment, tax exemptions and the like to those who settled in them. Six towns were included in this category: Ascalon, Beit She'an, Tiberias, Nazareth, Afula and Safad. In addition seventeen new towns were established: Eilat, Ofakim, Ashdod, Or Akiva, Beit Shemesh, Dimona, Hatzor, Javneh, Kfar Yeruham, Migdal Ha'emek, Ma'alot, Azata, Kiryat Gath, Kiryat Melachi, Kiryat Shmonah, Sderot and Shlomi. On 1 October 1960, these twenty-three towns had a population of 169,833 (36,528 families and 1,759 unmarried individuals).

Only 0·9 per cent of the inhabitants of these new towns were native Israelis; 81·2 per cent came from Africa and Asia, and 17·9 per cent from America and Europe. In the latter category the majority come from Eastern European countries, while only a very small number are from Western Europe and America. The ratio between these two classes of inhabitants of the new towns is 1 : 30.

We have already referred to a number of new towns in Israel in discussing certain regions of the country. In respect to size, economic and social character, these towns vary widely. The populations of the six older towns range from 10,000 to 25,000 inhabitants; five of the new towns have from 5,000 to 10,000 inhabitants and eleven from 1,700 to 5,000. Only one of the new towns has a population exceeding 12,000.

In a more advanced phase various categories of experts—town planners, civil engineers, sociologists and economists—began to voice their views on the new towns. The latter, of course, are open areas; any institution which wishes to do so can act within them freely through its representatives. Here activities have not been co-ordinated. Towns are growing but other towns are being deserted. A process of migration from the smaller to the larger towns can also be discerned, or the smaller towns are merely a place of residence for those of their inhabitants employed in the larger centres. A closer and more detailed study than has been made hitherto is necessary. Articles are being published in various journals on the dimensions of the development townships and on building styles; statistics are being published on the inhabitants, on housing, employment in industry, handicrafts, commerce, education and culture, administration, liberal professions and so on and so forth. Some of the topics which have been dealt with in greater detail in such studies are adult education, the reading of books and newspapers, entertainment and sport, political parties, social activities and juvenile delinquency.

But in the meantime no authoritative definition of the term 'development town' has been formulated. Those dealing with them from one aspect or another regard them as centres for which Government and public institutions have adopted a policy of earmarking resources to promote their development—hence their name.

New towns, needless to say, have been founded for many thousands of years. We can witness today in a large number of countries a similar process. The following extracts from an article[1] discussing one of the fifteen new towns established in Great Britain makes very interesting reading.

'To build a self-contained community of 60,000 souls in 20 years, and to keep them happy while you are doing it, is a herculean task.'

... 'Stevenage, like Britain's fourteen other new towns, still requires a mentor: this is the Stevenage Development Corporation (a non-elected body responsible to the Ministry of Housing and Local Government).'

... 'The Corporation controls the town's expansion by balancing the number of jobs it attracts with the number of dwellings it constructs. Firms moving to Stevenage are allotted houses, which the corporation finances with money borrowed directly from the Treasury, for the workers they bring and for further employees as they are hired. (Priority is given to qualified applicants from the London area).'

It is indeed a small world and there is close resemblance between

[1] 'Stevenage Growing Up'. *The Economist*, 5 August 1961, p. 545. London.

the problems encountered in Israel and in England—where the goal is wider distribution of the population by diverting inhabitants of the great cities to small towns. This is an aspect we must tackle too, even though Tel Aviv is a pigmy compared to centres like London and New York. A more important objective is the settlement of empty districts. New towns are being planned for the Negev with this end in view. We are discussing the development towns from another point of view. Where they are established in agricultural areas every effort must be made to integrate them economically and socially in their rural setting. Twelve, at least, of the seventeen towns we have referred to could be developed in keeping with the principles of *agrindus*. This is possible though we are fully aware, as in the Stevenage experience, that 'it is a herculean task'.

. . . 'At the beginning, employment and housing claimed priority. The early settlers lacked many amenities, and even necessities . . . It was too easy to place the onus for all personal ills on the town itself: if one disagreed with one's husband, if one's child was unco-operative (or had urges to rip cinema seats) it was all Stevenage's fault.'

But:

. . . 'The real fruition of Stevenage as a community now depends on its second generation, who will not be the rootless immigrants their parents once were.'

The author enlarges upon the petit bourgeois ideals that may be fulfilled and the amenities that will be enjoyed in the future Stevenage. There is no escape, it seems, from these ideals. The citizens of our own development towns wish to have the same facilities as have the citizens of Tel Aviv. We would like to aim higher. We want the younger generation of the development towns to mix and mingle and to merge with the young people of the *kibbutz*, in a co-operative environment. Neither co-operation nor even collectivism implies abstinence from the pleasures and amenities of this world. There has been an immense advance in the *kibbutzim* in housing, in cultural life and in living standards generally. But permeating all this is the realization of lofty social aspirations. We would wish the combination of industry and agriculture, of town and village, to lead to social integration. In this respect, indeed, the new towns in the development areas may differ from Stevenage. We should advise planners of new towns elsewhere to integrate the latter in their rural environment, in harmony with the *agrindus* idea. Stevenages on *agrindus* principles would help Britain provide for the fifty million new inhabitants it must expect in the next two generations; they would make room in the United States for a population increase of one hundred million;

180

they would create the framework to absorb the fourth billion of the world's inhabitants.

In the beginning the development towns were 'designed' for four to five thousand inhabitants. A little later sights were raised and the target set at 8,000, then at 12,000. Now, a town-planner has suggested

'a clear distinction between town and village within a regional context, on the lines proposed by Professor H. Halperin, is in harmony with the conclusions he has drawn in regard to the minimum population of 15,000–20,000, and even in regard to the maximum population of 35,000–40,000 in keeping with the economic structure and development of the region.' [1]

But the problems of the new towns of Israel are nevertheless different.

A Labour Ministry sociologist, within whose field the development towns come, has stated: [2]

'In the development towns new social relationships are being created, which like society in general are of a very complex character. The nature of society and of various ties in the traditional, fixed and stable society are totally different from that of a young society, composed of different ethnic groups, distinguished from each other by their culture, their traditions, the level of their social consciousness, and even their economic status. The location also constitutes a factor shaping the character of the society, which is different in an industrialized, urbanized, and in an agricultural context. And in touching upon this question of social relationships in new towns in the development areas, we must observe it from both aspects: on the one hand the negative aspect—that of the organization of the common collective interest of the dissatisfied, whose roots are still in the past, from which they have been plucked; but who have not succeeded in striking root in their new environment. The common denominator of the individuals in such a society is the force pulling them backwards and their common memories, towards their past which they have idealized.'

On the other hand there is the positive aspect: the motives binding a man to his new environment, promoting mutual intimacy by common activity. This is produced by a quasi-primary relationship, almost of identification, with the institutions, a recognition of common authority and prestige and in a later phase a proper distribution

[1] 'Dimensions of the Development Towns'. Yaacov Ben Sira. *Hahevra*, No. 8, August 1961, p. 21.
[2] 'New Towns in Development Areas'. Dr Alexander Berler. Davar's Housing and Construction Supplement, 1 July 1961.

of social functions, etc. In the majority of cases the two aspects are connected and integrated: as the residents become closer to each other and to their new place of residence, the external, negative, force is weakened.

In addition to the social conditons with which we have dealt only superficially, other factors of a more general character, such as the economic success of the new settlement, the period in which it was established, the lay-out, the availability of local initiative, also exercise their influence.

On these points Dr Berler says:

'The degree of economic success achieved by the new settlement exercises a direct influence upon the degree in which the inhabitants identify themselves with the locality. Under our conditions the diversion of resources by the State constitutes the first step towards economic consolidation. The circumstances dictating this diversion of resources are many and various—planning, employment, geography, etc.

But the period in which the settlement was established is of decisive importance. The foundation of a settlement under pressure of mass immigration did not permit the creation of an economic basis for employment. As a result of this uncontrolled phenomenon an inert, negative selective process was set in motion. The stronger residents endowed with initiative and means of their own left; thus the character of the settlement was fixed for a long time to come.'

. . . 'Another factor shaping social initiative is the composition of the population.'

The reference, of course, is to social environment, the presence of a nucleus of established residents capable of bringing the new immigrants closer to the way of life already developed by their predecessors. Not every established Israeli is capable of undertaking this task. An agricultural neighbourhood of a collective or co-operative character which is not closed to the inhabitants of the development town, but operates as an active factor and regards the education of the new immigrants as a national and social mission, is what is necessary. Special attention, of course, must be paid to the younger generation, to the children. The primary and high schools, their teachers and their pupils, the clubs and youth hostels, can also contribute towards establishing vital contacts with the environment, thereby accelerating the process of social absorption and the fusion, in the long run, of the diverse communities into a single national entity. All this, however, is not sufficient; the impact upon the budding new life will be weakened, if a comparatively large number of those upon

182

whom the newcomers could model themselves, to whom they could look for leadership—the teachers,. the doctor, instructors, public leaders, the journalist—do not take up permanent residence in the settlement, or if they regard it as no more than a place of employment, from which they return home when the day's work is done.

In Israel conditions differ radically from those obtaining in other countries, in Europe, Australia and America, where new towns are also being established. In an article entitled 'Communities in the Making', published in the London *Economist*,[1] we read of complaints about 'the absence of the friendliness of city life'. But the author adds, 'Some of the criticisms of all newly built communities (the new towns included) really boil down to criticisms of modern life'.

This cannot be said of Israel. Here the newcomers from Africa and Asia who make up the population of the new towns are many centuries behind 'modern life'.

'England is changing everywhere, but the changes are sharper and quicker in the new towns. These towns are in fact a forcing ground for middle-class values, and it is rather silly to try to put back the social clock by encouraging there a kind of bogus urban matiness. Yet this is just what the Ministry of Housing seems now to be proposing. It suggests that new towns might become more "neighbourly".'

In England the remedy is being sought in housing design. Indeed the article quoted is entirely devoted to this subject. This is also the case in Israel, and memoranda, studies and articles on this subject have been published in profusion. This, regrettably, is not our only concern in regard to the new towns.

We have the highest regard for the devoted concern shown by the various Ministries, Government and public institutions to this aspect of settlement in the new towns. It is interesting to note that the Ministry of Housing has set up a department for public institutions in the new towns also for this purpose. Special attention is paid in England to housing, for which, indeed, a special Ministry has been set up whose functions go far beyond the technical aspects. In Israel the Ministry of Housing has established a committee to co-ordinate its own activities with the Employment Authority and the competent departments of the Jewish Agency engaged in bringing immigrants to this country and in making the initial arrangements for settlement on the land or in the development towns. This co-ordinating committee is in close contact with other public bodies operating in these areas, such as the Ministry of Trade and Industry's department for promoting industrial enterprise in the development towns, with the Ministry of

[1] 2 September 1961, p. 869.

Interior, which is ultimately responsible for municipal organization with the Amidar Housing Corporation, with the Ministry of Education and Culture and the Absorption Department of the Histadrut (Federation of Labour). Research in this field comes within the scope of the Authority's sociologist, who, needless to say, collaborates closely with the committee.

This central committe in turn has set up local bodies with a wider representation, its members coming from the local authority, the Labour Council, the labour exchange and indeed any other public body which is active.

Between the central and the local committees there are district committees. The procedure adopted is as follows. The local committee studies all matters brought to its notice, decides upon the priority to be given, makes its recommendations and remains in close touch when and in so far as the latter are carried out.

In sum, what is achieved is a streamlined budget upon which all the competent bodies operating in the development towns have agreed.

Research is conducted in the spheres of planning and construction, and in the development of social relations in these localities. Social studies have so far been made in the following fields: demographic structure; social background; economic conditions; educational and cultural institutions; youth and the family, study and work; public activity, etc. The methods used are ecological and typically biographical, and are based upon individual study. Every effort is made to adapt the finding to local conditions.

The annual reports which have accumulated render a picture of the homogeneity or heterogeneity of the population of the new towns and provide important data for consideration of internal relations— particularly in the direction of the friendliness and neighbourliness discussed in the *Economist* article referred to. The reports indicate the rôle of each ethnic or age-group in consolidating local society; when such consolidation is not regarded as a likely prospect and the social framework is in jeopardy it may be decided that dissolution is the only way out. These reports also indicate the degree to which public officials have become integrated in the local community; the impact and importance of such integration; whether, as the case may be, they should be changed or required to take up permanent residence in the area.

On the basis of the data collected the development towns can be classified in three categories. In the first come six older towns— Ascalon, Tiberias, Javneh, Nazareth, Afula and Safad—where the object of development was to overcome ingrained backwardness and to rehabilitate the towns, to expand them and to construct new quarters, to promote the establishment of industries including the transfer of existing factories from other centres, in order to achieve a wider distribution of the country's population. In addition they were to serve as regional centres. Such development will raise them from their present dimensions to the status of medium-sized, go-ahead towns. Close to them in certain respects are the new towns, for which no rôle of serving as regional centres can be envisaged, at least not for the time being. This category includes Eilat and Dimona and perhaps also other towns in the Negev scheduled to rise at some date in the future. The purpose of the latter, principally, will be to populate the empty wastelands. For this reason indeed every possible aid must be extended to them in the establishment—and the transfer from other areas—of factories. The location of the towns in the third category (including the ancient town of Beit She'an) places them in a special class. At present they are dependent upon their agricultural environment as the main—and often only—source of employment. Development may proceed upon one of two lines here. The inhabitants may continue to work in the agricultural settlements, as farm labourers, in building, services and industry. Economic enterprises, mainly of an industrial character, may be established within their own municipal area. At the time of writing we learnt from the press of a shortage of labour in Upper Galilee. The apple-picking season had absorbed most of the farm labourers (including women) in the town of Kiryat Shmona, causing an acute shortage of cotton pickers. The report assumed that early conclusion of the apple harvest would free 1,600 pickers to handle the cotton crop. To round out the picture, however, it should be mentioned that there is seasonal unemployment in the winter and relief work has to be provided for 750 men. Such sharp fluctuations in the labour market so characteristic of Kiryat Shmona will disappear, the experts aver, within three years when the orange groves in the district come into full bearing.

This situation, however, is typical of most areas. But having continued for twelve years, what guarantee have we that it will come to an end within another three? The population of Kiryat Shmona—and of this country, and the world in general—will grow considerably within the next three years and we can by no means be sure that we shall no longer have the problem of labour fluctuations on our hands.

But there is an alternative course that may be followed. Towns like these, in advanced agricultural areas and dependent economically upon the villages in the region, can assume an entirely different character if they are fully integrated—economically, socially and in every other respect—with their environment. The transition from the existing situation to that which we have outlined may be extremely difficult. In certain cases it may even prove completely impossible.

185

We have already noted above that in Utah, when a rubber factory was established in a rural area, some of the farmers and some of the new factory hands became part-time farmers. We do not regard this as a desirable process in Utah; we are strongly opposed to anything similar in our own system of regional co-operation.

There is abundant proof of the difficulty of converting an ordinary urban settlement into a regional centre for an agricultural hinterland. It is many times more difficult when that hinterland is of a co-operative character; viz. settled by *kibbutzim* and *moshavim*.

In the distant past, ancient Tiberias might have served as some sort of provincial capital for the Jordan Valley. It did not regain that status when the Valley was resettled by *kibbutzim*. On the contrary, it remained in its age-old lethargy. Zemah, the Arab town which was nearer geographically than Tiberias to the new *kibbutzim*, no longer exists.

Beersheba in the past was the provincial capital of the Negev when the latter was inhabited sparsely by nomad Beduin tribes. To a certain extent, from the administrative, transport and economic aspects, it retains this position, but it is not the capital of the new Negev, that is the Negev which today is dotted with *kibbutzim* and *moshavim*. Socially it is a town unto itself. Nor, to cite another example, has Afula, founded only in 1925 and located in the geographical centre of co-operative agricultural Jezreel Valley, achieved the status of the regional centre. Its founders, it is true, called it 'City of Jezreel', envisaging its development into the centre of the large and important district in which it was situated. But after thirty-seven years it is still hardly more than a road junction. Why? The only reason is—lack of planning. When it was suggested that a sugar refinery or flourmill or feed mixing plant be established there jointly by the settlements of the district, *kibbutz* leaders declared that 'co-operation between communes' could not possibly serve as a substitute for the idea of 'a commune of communes'. Some of the *kibbutzim* of the Valley, which have recently celebrated their fortieth anniversaries, have been riven in two three times already—at the beginning of the twenties, in the middle of the twenties and again in the middle of the fifties. They are as far as ever they were from 'the commune of communes' but in the meantime they have lost golden opportunities of furthering regional collaboration on a co-operative basis.

Development towns of the first category we have outlined cannot serve as centres for the districts upon which they are dependent; their inhabitants are, socially and culturally, hundreds of years behind their environment. At best they are residential quarters of the wage labourers employed in the neighbouring settlements.

186

Elsewhere we have discussed the keen and highly interesting ideological ferment which is going on in the Sha'ar Hanegev region. The regional town of Sha'ar Hanegev is Sderot, founded in 1951, as a workers' quarter. The newcomers were allocated allotments upon which to develop agricultural small-holdings. When the population grew, Sderot became a town and was no longer content with seasonal employment. Later it was officially recognized as a local council and withdrew from the regional council, which represents ten *kibbutzim*, one *moshav* and two farming estates. Only Sderot found their company uncongenial.

Sderot is constantly absorbing new immigrants, and in 1960 it had a population of about 3,500, but the element possessed of drive, initiative and ability had left. Eighty-eight per cent of the inhabitants come from North Africa, 10 per cent are from Egypt, Persia and Iraq, while only 2 per cent are of European origin. Families here are large and immigrants coming from the North African and Middle Eastern countries have customs and traditions totally different from those of the European Jews or of the Israelis. They maintain the extended family structure of *hamulot* (clans) between which a powerful antagonism still exists, so much so indeed that intermarriage between *hamulot* is rare. In the State of Israel measures are not adopted to undermine established customs, but this custom of closed leadership, which hampers development and is one reason for the backwardness of the younger generation, must be abolished. If it were only possible to insist that public officials and representatives live in the town they could have a fruitful civilizing influence.

It is significant that in the town of Sderot there is not a single shop, or even stall, for the sale of newspapers—Hebrew or foreign. The ratio of radios is low—one to every ten families. There is not a single organization, or any other evidence of the public initiative of the local inhabitants. Even the sporting club which has a comparatively large membership was established and is administered by non-residents. Every office or public institution—the Labour Federation and its constituent bodies, the Workers' Bank, the Dues Bureau, the clinic, provident and insurance funds, the Working Mothers' Organization, the Working Youth Association, Mishan Welfare Fund —all are conducted by non-residents.

The situation is hardly different in other development towns, though there may be local nuances.

Sderot as it is today cannot possibly serve as a regional centre; it cannot be the location of an *agrindus*. Even Kiryat Shmona cannot undertake such a function, let alone other new towns in organized agricultural areas.

187

A solution, we believe, may be found in the near future by establishing the nuclei of industrial enterprise and cultural activity of the *agrindus* within the development towns. Local residents who are employed by the *agrindus* must be organized in the proper societies. Such a method could have a far-reaching educational effect. Members of the *agrindus* as far as possible should fill the various public positions in the town, and gradually educate the local residents to undertake public responsibility. This transitional stage may be of prolonged duration—as much as a generation or even two—and must constantly be kept on its course towards the development of an *agrindus*.

Elsewhere we have pointed out that in the urban centre of the ideal *agrindus* there need hardly be any residential quarters. It will be the location of the factories, the workshops, the services, the power station, the waterworks, the post and telegraphs office, the telephone exchange, the offices, the schools, the amphitheatre, cinema and sports grounds. There will always be people there in the factories cleaning up, watchmen, waiters in the restaurant, etc. The workers will live in the agricultural settlements which maintain the *agrindus*. There will also be a type of *agrindus* the members of which do not belong to any village but to co-operative societies for production and services. They, of course, may live in residential quarters in the urban centre.

The recent organization of thirty-four co-operative groups of boys and girls can serve as an instructive example. The groups began to take shape during study-days organized by branches of the Vocational Section of the Noar Oved (Working Youth Association), in the course of lectures given at trades schools, in seminars for secretaries of trade union sectors and for teachers of trades schools. As these groups assumed a more crystallized character, instruction and explanatory activity among them was begun, directed towards methods of establishing co-operative enterprises. Regional meetings of the members were also held. Work on all levels, aiming at strengthening the vocational and social organization of the existing groups and the creation of new ones, continued all the time.

The young people are sixteen to seventeen years of age, and the military authorities have agreed, in so far as this is possible, to employ them in their diverse trades during their term of military service. Thus service in the Army, far from interrupting their training, will promote it.

It is noteworthy that these groups have been organized in the development towns as well as in the larger cities, and also in Arab Nazareth. The regional gatherings serve to strengthen mutual relations between the groups and to create a suitable social atmosphere.

The publication of an organ, 'Yad Beyad' (Hand in Hand), is under consideration.

Once the group has taken definite shape and has been approved by the local branch of the Noar Oved its members elect a committee, secretary and treasurer.

Concurrently the nature and importance of co-operation, the contribution it can make in furthering the future career of the young person and in ensuring suitable employment after the conclusion of military service, are explained to the parents.

Should this movement expand and strike root among the young people it can improve the prospects of the industrial centre of *agrindus* which will be mainly of a co-operative character.

The development towns, in areas which seem most suitable for the establishment of *agrindi*, are *faits accomplis*. The *agrindi* that will be set up will adjust themselves to these facts when most of the local inhabitants will be employed in the *agrindus* as members of the constituent societies and, in the transitional stages, as hired employees.

The prospects for the establishment of the ideal *agrindus* are today meagre. As stated, suitable conditions still exist in the Jordan Valley and in the Hefer Valley, and perhaps also in a number of other regions where development towns have not yet been established.

XVII

SUMMARY AND CONCLUSIONS

PREMISES FOR THE ECONOMIC PLANNING OF AGRINDUS

(1) Within the *agrindus*, as within every other framework of human activity, social, economic and political factors will operate. Any attempt to divide it up into its components inevitably prevents us from seeing it in the round, as an entity. Nevertheless we shall now analyse the diverse economic forces and considerations promoting the creation and development of the *agrindus*, as we have concentrated throughout this volume mainly on the social aspect. Isolation of the economic from the socio-political factors will enable us to foresee certain problems of the economic processes of the *agrindus* and to outline possible solutions. However, in every stage of our analysis we must constantly bear in mind the operation and the importance of the social forces in shaping the solutions we have to offer. Thereby we can ensure that our suggestions will be of a harmonious and therefore practical character.

(2) In the broadest terms the economic process within the *agrindus* includes the following activities:

(*a*) Primary production within the agricultural settlements.

(*b*) Secondary production both in the settlements and in the urban elements of the *agrindus*. In this stage agricultural produce is processed.

(*c*) Productive activities serving the main production processes carried on both in the settlements and in the urban elements. These activities are ancillary to the primary and secondary processes of production.

(*d*) Consumption services which may be located in all residential sections of the *agrindus*. These include distribution and supply activities; medicine and hygiene; education, culture and the like.

(*e*) Transport services to carry men and materials between the

190

various elements of *agrindus*, and also from the *agrindus* elsewhere.

As pointed out these various activities take place in different locations. They are interdependent through the flow of goods and services between them and through their competition for the limited resources available within the *agrindus*.

(3) In analysis and solution of economic problems, in isolation, it will be reasonable to assume that the material well-being of the population of the area is a monotone increasing function of the income of the *agrindus* as a whole. For this purpose we may define '*agrindus* income', as analogous to 'net national product', as follows:

Agrindus income:

  = Consumption (private and public)[1]
    + Net investment in *agrindus*[2]
    + Net external investment[3]
  = Gross value of output
    — the value of intermediary products and services and improved inputs consumed in the process of production.[4]

We shall not enter into the question of the distribution of income among the inhabitants of *agrindus* at this stage. This is dependent upon

  (I) Distribution of ownership of means of production.
  (II) The remuneration for various means of production.
  (III) Social factors.

In the main the weight of these three factors will be determined in the non-economic sphere of the *agrindus*.

Our assumption concerning the correlation between the economic well-being of the inhabitants of *agrindus* and the '*agrindus* income' provides us with a criterion to examine economic decisions. In other words, within the framework of social conditions prevailing in the *agrindus*, and given the restrictions on available resources, we shall always prefer the economic alternative associated with maximum '*agrindus* income'.

(4) What are the economic problems confronting the planner?

---

[1] Calculated according to national market prices plus (or minus) costs of transport depending upon whether such goods or services are imported into or exported by the *agrindus*.

[2] Net investment = gross consumption minus capital consumption.

[3] Net external investment is equal to the surplus in the *agrindus* balance of payments.

[4] Including capital consumption and interest payments to outside investors.

SUMMARY AND CONCLUSIONS

(a) *Problem of planning data*
Economic planning does not engage in creation *ex nihilo*. Economic facts exist and come into being everywhere. These constitute the basic data of planning. Moreover, in the more comprehensive economic life of which *agrindus* is a part, constant changes are being registered in respect of demand, technology, international trade and so on. It is the task of the planner to estimate the relative weight of each of these factors. His primary problem, accordingly, is to collect the relevant data governing present conditions and possible trends in the future. In addition, of course, he must be in possession of all the facts relating to the diverse activities we have already enumerated.

(b) *Choice of activities*
The problem posed is one of choice of activities capable of maximizing the income of the *agrindus*. We must bear in mind the mutual connection between these activities. In other words choice of one activity may frequently involve choice of another.

(c) *Level of activities*
The level of various activities is, of course, dependent on available resources. Restriction of resources should therefore be accounted for by the planner. Because of interdependence of a relation, the levels of some activities are often determined by the levels of others. Some activities, such as services, ordinarily depend on the size and composition of the population, its cultural character and the level of disposable personal income.[1] The last, of course, is a function of the productive operations and the investment policy.

Various programming methods including linear programming are at the disposal of the planner in determining the choice and level of activities. In so far as the levels of productive activities are given and their mutual relation is known, the Leontief input-output analysis can be utilized in order to determine the level of *agrindus* activities generally, the demand of the branches for factors of production and the volume of general output. The consumption functions of services and goods—viz. the relation between consumption of goods and services and disposable income—can be calculated and integrated in a model.

(d) *Returns to scale*
In this context the problem of determining the values of the interdependence coefficients to be used in the analysis arises. In many economic activities, per unit costs of production depend on the level of output, i.e. these activities are characterized by various degrees of

[1] The disposable personal income is made up of the income of the *agrindus* less the value of public consumption and less the value of income not distributed but directly invested in the enterprises.

192

returns to scale. In accepted linear programming and input-output models such economies of large scale production are ordinarily assumed away.

Thus, in one way or another, these models must be integrated in a study of returns to scale of various activities.

The problem of returns to scale is also important for determination of the number and location of enterprises. Again we shall always give preference to the plant maximizing the *agrindus* income as defined above. An increased scale of plants may lead to a reduction in per unit cost of production. If, however, raw material is to be procured or output be distributed over a wide geographical area, an increase in scale may imply higher transportation costs. Limited demand for the plant output may place another restriction on plant size. If a considerable increase in costs is involved, it might be desirable to integrate the particular plant within a broader framework.

(e) The location problem

The problem of the location of the economic operations constitutes the axis about which the *agrindus* economy revolves. In the final analysis the location will be determined not only by economic factors; the social factors are also of major importance. The former, however, will always weigh heavily in such a decision.

In the preliminary stages of the formation of an *agrindus* the question will arise of the location of projected agricultural settlements and urban centres. In more advanced stages this problem will cease to exist, but there will always be the question: Where should new plants be sited, in the agricultural settlements or in the urban centres, and in which of them? Such factors as procurement of produce and transport of workers, despatch to markets, provisioning of settlements, must all be taken into account. In combination with returns to scale and optimal levels of various operations, economic considerations will dictate the location of both productive and service activities.

In this respect, however, planners have not got tried and tested methods at their disposal; these must yet be developed. The Location Theory as developed by Isard, Lösch, Hoover and others can serve as a guide in the formulation and application of the requisite planning methods.

(f) Optimal development of the 'agrindus'

There are two sources for financing the development of the *agrindus* economy:

1. internal accumulation of capital
2. outside funds.

Utilization of the latter source will be governed by rates of interest and terms of repayment on the one hand and on the other by the marginal efficiency of investments in the *agrindus*. How such funds are raised and what recourse the various elements in the *agrindus* have to external sources does not come within the scope of the present study; it belongs properly to the socio-economic sphere.

The degree of internal accumulation of capital within the *agrindus* is a function of its private and public propensity to consume. The administrative stratum[1] can influence consumption patterns by its policy in regard to public consumption and private incomes. By restricting public consumption and reducing that part of the income distributed to individuals in the family villages and in the urban centres, or diverted to consumption in the collective villages, the process of capital accumulation can be accelerated or decelerated. Decisions on this issue come within the socio-economic sphere and lie beyond the boundaries of our study.

The question of allocation of capital to the various operations is of course primarily of an economic nature. In this respect what is important is not only the general scheme of allocation but also the timing of the investments made. The goal must be to time the investments in such a fashion as to ensure a growing income for the *agrindus*.

To solve the problem posed by allocation of capital and timing of investments planners can adopt dynamic linear programming. This method, it is true, has not yet been properly tried out, but it seems to be the most suitable in solving problems of this kind.

(5) *Vertical integration.* There is no practical limit to vertical integration. The question, to what extent we can go in this direction in our *agrindus*, is still open. Planners, no doubt, will have to compute interdependence coefficients between our own restricted sector and other economic sectors and regions in the state. In the Israel context there is the very topical question of enterprises for the processing of agricultural produce, operated by two, three or even more *agrindi*. In other words, it will be a question of inter-regional co-operation after which, we have no doubt, that of inter-sector activities will be raised.

The study and planning of *agrindus* make up a complex of problems most of which we have already isolated in previous sections. In respect of a number of problems there exist methods of planning and analysis which can be noted comparatively easily. On the other hand, however, there are a number of problems for which adequate methods have not yet been developed. This is a subject of very wide scope. The problems encountered are closely inter-related. It will prove

[1] This term includes all the personnel and institutions of *agrindus* taking the major decisions affecting economic process.

necessary to devise methods to deal with specific problems and also some means of integrating these partial solutions into an articulate whole. This, of course, is a task of formidable proportions. It presents a challenge to all those interested in economics and economic planning.

'AGRINDUS'—AN ECONOMIC CATALYST

The purpose of the present essay has been no more than to introduce and to give a broad outline of the idea of *agrindus*. At this stage we have preferred to refrain from developing any formulae. We have sufficient empirical material at our disposal upon which we could base such formulae: if we have not done so it is because the examples brought from typical regions in Israel, while tending towards agro-industrial integration, still have a long road to go before they can breathe the idea of regional co-operation into the process of industrialization in agricultural districts. What we have rendered in the foregoing has been no more than the bare skeleton of what we propose. The method of application that has to be developed must be adapted to the special—national, intellectual, social, cultural and economic—conditions obtaining in any given location or state.

'The real duty of the economist,' we agree, 'is not to explain our sorry reality but to improve it.' [1] Here we have endeavoured to point out a path leading to improvement.

As long as *agrindus* is no more than an idea, until it has been translated into reality, the critics can offer their animadversions at will. They may be right, and what the present writer has outlined may be condemned as utopian. Thomas More's ideas, of course, have not been put into practice because they were not practical, but for four hundred years and more Utopia has been the lighthouse of visionaries and dreamers. Many social theories, which in the course of generations have been incorporated in social reality, first saw light under the impact of Utopia. The very fact that it has been constantly recalled since the sixteenth century, whenever any new idea, any social or political reform, has been mooted, points to its educational significance. It may be in the nature of a paradox when we say we recognize its *practical* value, not in itself, but as a stimulant to vision which has become reality. In Israel, indeed, within the short space of two generations, ideas that seemed at the time fantastic and were written off as 'utopian' have taken on flesh and sinew.

And even should an *agrindus* be eventually established it may be regarded as an isolated case of no wider significance. The theory

[1] August Lösch, *The Economics of Location*, Yale University Press, 1954, p. 4.

195

behind it may be criticized as schematic, certainly not on a par with the ideas which J. H. von Thünen developed in his 'Der Isolierte Staat in Beziehung auf Landwirtschaft und National Ekonomie' (Hamburg 1826). The present study does not pretend to compare with von Thünen's work of genius, with the richness of his fantasy and the clarity with which he presents his ideas. 'The Isolated State', of course, laid the foundations of a new doctrine of economics of location, which is steadily becoming an important subject of study of considerable practical value.

We have been led to the *agrindus* idea not only because we have seen that the advantages of a large tractor of the D8 type have induced neighbouring villages to combine to purchase and operate one. For the tractor has been followed by the feed mixing plant, the cold storage plant, the packing house and so on. The frontiers of co-operation have been pressed back to include a joint amphitheatre, school, closer social contacts. From two Danish neighbours owning one plough, two farmers sharing a single horse, a number of farms jointly operating a tractor, the economies of scale led naturally to broader and broader partnerships. In an advanced society, like the collective settlement in Israel, partnership develops into a steadily expanding co-operation.

In Soviet Russia, in the course of the thirty-three years of their existence, both the *kolkhoz* (collective farm) and the *sovkhoz* (Soviet farm) have known many changes—in the dimensions of the farms, the conditions of operation, remuneration, etc. More recently increasing stress is being put on the connection between agriculture and industry. In one article we read as follows:

'In discussing the extensive potentialities of the *kolkhoz* in promoting production we must stress emphatically that the technical basis can only be changed by the development of State industry . . . The tasks confronting State industry in ensuring technical plans for the expansion of agricultural production are many and various. The *sovkhozy* have a highly important part to play in the perfection of the collective system of farming. They must serve as an example of rapid progress in labour efficiency, in reducing costs of production, in increasing output by specialization. The Party must make every effort to overcome the backwardness of the economically weak *kolkhozy* . . . Within the next decade output per worker must be doubled, and within the next twenty years, quadrupled . . .

The processes of mechanization, electrification and chemicalization of production, are steadily advancing in both the *kolkhozy* and the *sovkhozy* . . . The object is to enable every man and woman in the *kolkhoz* to adopt an occupation of his or her own choosing, to

196

specialize in their field, to demonstrate initiative and to work enthusiastically and in a creative spirit.

The programme of the Party declares that as part of the development of the *kolkhozy* and the *sovkhozy* the productive ties between them, and also with local industrial enterprises, must be reinforced, and practical experience must be expanded for the joint organization of industrial plants. Thus more continuous and effective utilization of the labour force and productive resources will be ensured throughout the year, labour efficiency will be improved and the material and cultural standards of the population will be raised. Gradually agro-industrial societies will be created to promote economic advancement. Into these societies the agricultural farms will be organically integrated with the industrial processing of their output, by way of rational specialization and co-operation between agricultural and industrial enterprises.' [1]

Over the past two years there has been a distinct and growing trend towards inter-*kolkhoz* collaboration on a regional basis.

'Life shows that the integration of agriculture and industry in a single, unified process, conditioned primarily by the needs of agriculture, holds out a perspective of objective economic development', one Soviet economist writes.

In certain areas this agro-industrial integration is receiving a new impetus. In Kaushan, in the Moldavian Soviet Socialist Republic, a general meeting of representatives of the *kolkhozy* resolved to set up a regional inter-*kolkhoz* union, which was to comprise an inter-*kolkhoz* institution for building or for the manufacture of building materials, mechanical maintenance workshops, a productive combine (mills, oil refinery, feed mixing plants, bakeries, fruit, grapes and vegetable processing plants, etc.), institutions for utilization of surpluses, inter-*kolkhoz* artificial insemination stations, supply stores, plants for the manufacture of antibiotics, schools for the training of specialists, a clinic, a youth camp, a resthouse, a mutual aid fund, etc.

The limits of inter-*kolkhoz* industrial plants cannot be invariable. They change and continue to change in keeping with the development of industrial and agricultural production with processes of greater division of labour between them and advancing specialization.

... 'As a result ... institutions will be organized, rooted in agriculture and industry on a co-operative basis, integrating agriculture with the industrial processing of its products.

[1] 'Ekonomika Selskogo Chosiaistva'. Editorial. The Collective Farm is the school of communism for peasantry, Sept. 1961, pp. 7–10.

. . . There will be basic changes in the distribution of the population throughout the country.

. . . The creation of these united agro-industrial plants will be accompanied by changes in the smaller settlements, which will be converted into rural-urban settlements combining the comforts of the city with the advantages of the village.'[1]

From the little information we have on People's China we learn that they too are endeavouring to emulate the example of Soviet Russia.

Over three decades Soviet Russia has endeavoured to reorganize and to change methods of conducting its agriculture. The social trend is persistently towards collectivization.

In the United States, however, scientists put the accent on 'the intensive man'. Several very interesting works have been devoted to the problem of 'capital formation by education', of 'human capital', etc. In this context they believe that 'human resources' weigh more even than 'natural resources' (land).

As the protagonist of the idea of *agrindus* is the intensive man who integrates agriculture and industry, town and village, mainly by co-operation between men and societies, we have a special interest in studies of 'economic capabilities of people' which are in full harmony with our present discussion of their social capabilities.

. . . 'Natural resources are not among the more promising sources because they have become so small a contributor to national income (only about 5 per cent of the national income of the U.S. comes from this source) and because farm land bulks large and good farm land is no longer around for the taking. Of course, there is some land that can be cleared, drained, or irrigated, but surely the Dutch are paying a high price for the little economic growth they are getting from driving back the Zuider Zee, and the Italians from developing a little farm land, taking it from the Adriatic Sea at the mouth of the Po. More irrigation in India or in California, even though the Federal Government faces the bill, is not a cheap source of additional income . . .'

. . . 'The principal source of income is from labour . . . In the United States, natural resources contribute approximately 5 per cent, reproducible capital goods 20 per cent, and labour 75 per cent of the national income. But what is only vaguely understood is the fact that the *capabilities of labour are a produced means of production.*'

. . . 'The principal activities that produce human capital are education, on-the-job training, and health services' . . .

[1] Voprosi Ekonomiki N. 10, 1961, pp. 83–91. Voronin, 'O Sotshetanii selskokhosiastvennogo i promishlennogo proisvodtstva v derevnie'.

SUMMARY AND CONCLUSIONS

... 'Between 1929 and 1957, the real income of the U.S.A. rose to $302 from $150 billions, an increase of $152 billions. Additional education in the labour force contributed, according to this estimate, almost $32 billions, representing 21 per cent of the total rise in national income' ...

... 'Even this low estimate implies that education has been contributing as much to economic growth as has the increase in the stock of conventional capital goods' ...

... 'Low-income countries would, of course, like to acquire additional income streams as cheaply as possible' ...

... 'There are indeed strong reasons for believing that most low-income countries are under-investing in human capital' ...

... 'India, notwithstanding her many efforts to increase food production, is not doing well, whereas Japan with two to three times as dense a population and relatively much less land for farming is expanding her food output rapidly ... Similar perplexities are to be seen in Latin America and throughout Europe, including Israel. Those countries that have the best records have not won them because they have the best land, or mainly because they have increased rapidly the reproducible capital in agriculture, but largely because they have acquired the skills and knowledge required, and a confidence in their ability to develop a modern agriculture. These capabilities would appear to earn high dividends at home and abroad under present circumstances.... A capable people are the key to the abundance of a modern economy. They are a major source of economic growth.' [1]

It is not only the United States that is improving its economic capabilities; the low income countries, too, as Professor Schultz witnesses in his paper, are doing their best in this direction. India, for example, despite the exceptionally difficult conditions prevailing there—she has the second largest population in the world—has made a remarkable effort in sponsoring new trends in her social and economic development.

In recent years, it may be noted, immense progress has been registered in her agriculture though yields are still among the lowest in the world.

On 2 October 1952, the Government of India proclaimed the inauguration of fifty-five community projects in districts where conditions were favourable in point of climate, soil and irrigation installations. Social development is designed, *inter alia*, to stimulate

[1] Theodore W. Schultz, 'U.S. endeavours to assist low-income countries improve economic capabilities of their people', *Journal of Farm Economics*, Proceedings issue, Vol. XLIII, Dec. 1961, N. 5, pp. 1068–77.

co-operation and to raise the cultural levels of the rural population.

The organizational unit for the purposes of this new plan was a region comprising one hundred villages with 60,000–70,000 inhabitants on an area of 150 square miles. This district was divided into groups of 5–10 villages. The operation came under the Ministry for Community Projects and Co-operation which was set up in 1958. This Ministry comes under the general direction of Indian Planning Centre which is headed by the Indian Prime Minister.

Thus it will be seen India, too, is seeking socio-economic methods to solve its most vital problem.

A different situation, which also cries out for a solution, is set forth in the Report of the F.A.O. for European countries.

'Farming [this Report declares] is becoming . . . more and more like a manufacturing industry . . . During the nineteen-fifties, when gross agricultural output in this region was rising by 27 per cent, the volume of labour input appears to have declined by somewhat more than 20 per cent. Although agricultural employment is falling, more labour is being employed in industry, so to speak, on behalf of agriculture . . . The process is, of course, from the long-term point of view, a gradual one which has characterized European agriculture to a greater or lesser extent for several decades, and which will certainly continue for decades to come.' [1]

The report confines itself to the economic aspect and reveals the general trend towards the capitalization of European agriculture. It ignores totally the social and political aspects. Further studies and views on these important questions may be expected.

We have already had occasion to refer to the I.L.O.'s study, *Why Labour Leaves the Land*. The concluding passage of the final chapter has some relevance in the present context.

'There is enough evidence to suggest that the effects of migration into cities involve social stress and tension on a scale unknown in the advanced countries. . . . In the chiefly agricultural continents movement to the city may mean a complete uprooting and a break with family, neighbours, custom and even religion. The gulf between town and country may be a gulf between two worlds.

. . . One common habit of the thought in regard to these social stresses is to accept them as inevitable consequences of the new "industrial revolution".' . . .

[1] Source: United Nations Food and Agriculture Organization. Towards a capital intensive agriculture. Fourth Report on Output, Expenses and Income of Agriculture in European Countries. Part I: General Review, Part II: Review of individual countries. Geneva 1961.

SUMMARY AND CONCLUSIONS

. . . 'Social stress is most evident in alternating movement, aggravating the wastage of manpower which this type of migration entails. It should be emphasized that such movement, where it is a chronic and not a transient condition, gives rise to the most serious problems of any considered in this study, and that its existence is in no sense a necessity of industrial development. It arises from the failure of agriculture to provide a livelihood and the failure of industry to provide adequate living and working conditions for its employees. It is therefore, in essence, a social problem which can be solved by measures of social reform.

In addition to the general necessity for balanced development, measures are needed to assist the underprivileged groups within the agricultural community. Of these, agrarian reform is the most important, accompanied by the organization of co-operative or community development, aimed at increasing investment and employment and providing a stronger social and economic basis for the life of the countryside.' [1]

We are familiar with the growing problems in the agriculture of the richest country in the world. The following opinions serve to complete the picture:

. . . 'The structure of American agriculture will continue to change . . . Vertically integrated operations will dominate the poultry industry . . . Large scale of factory-type operations will become increasingly important in livestock-feeding operations . . . Farm operators will become increasingly dependent on non-farm sectors of the economy . . . More and more farm people will need to turn to non-farm pursuits for employment . . . The proportionate numbers of farmers requiring off-farm employment will vary between types of farms in accordance with required changes in the organization and structure of their farms . . .

Factors limiting desirable changes in the organization and structure of farms by 1975 will include risk and uncertainty, managerial requirements and capital requirements, all of which will increase greatly.' [2]

But should not social and organizational solutions be sought as well?

[1] International Labour Office. *Why Labour Leaves the Land*. A comparative study of the Movement of Labour out of Agriculture. Geneva, 1960. pp. 227–229.

[2] H. L. Stewart, 'The Organization and Structure of Some Representative Farms in 1975'. *Journal of Farm Economics*, Proceedings Dec. 1960, pp. 1378–79.

A recognized authority on agricultural economics, discussing 'The Dimension of the Farm Problem', has stated:

... 'At this time I will only reiterate the point of view taken—that the farm problem is basically a problem of adjustment to changing conditions which necessitate an absolute decline in farm employment if the real returns to farm labour are to increase as rapidly as such returns increase in the rest of the economy ...' [1]

Throughout the entire world we see agriculture grappling with complex problems, which vary with location and local conditions. A vast amount of intellectual energy is being invested in all countries in this quest for a solution.

In this work we have engaged in an exploration of one path leading off from the main socio-economic highway. Perhaps we have progressed uncertainly and too rapidly. In self-justification we may recall a passage from August Lösch's introduction to his own study.

'No man-made system can avoid abitrary. None is conclusive for we do not know the final cause of things. We know only a mutual relationship, not a simple series of causes from beginning to end, from above to below.

It is basically immaterial where our presentation begins, since we cannot linger over the parts without considering the whole. If a system is regarded as an order of preference, the emphasis lies heavily on the word ORDER. We ORDER our facts according to viewpoints that are important to us. Hence the same thing appears again and again but is seen differently, whereas in a proper sequence, in an ideal system, it would have a unique place.' [2]

Agricultural development the world over is confronted with many grave problems. We are fearful for what fate holds in store for those social achievements of a special character which have been registered in Israel, at cost of so much pioneering effort on the part of two generations. We are seeking some means whereby to reinforce what has been accomplished. All mankind stands in need of the strengthening of the village as a national, social and economic category.

Perhaps the *agrindus* can make some contribution towards this objective in our own and the succeeding generation. The point of departure in the present volume has been co-operative agriculture in Israel, but we believe that the *agrindus* holds out a prospect for the

[1] Professor D. Gale Johnson, 'The Dimension of the Farm Problem', *Problems and Policies of American Agriculture*, Iowa State University Press, 1960, pp. 47–62.

[2] August Lösch, *The Economics of Location*, Yale University Press, 1954, p. 3.

solution of the problems of agriculture and the village in other parts of the world as well.

In our concluding chapter we have quoted the views of a number of authorities on agricultural policy and economy. These views reflect changes that are being registered rapidly in world agriculture and prove that the concern regarding the future course of development is fully justified. It is the task of our generation to endeavour to ensure a better future. Finally, we have underlined the economic aspect.

In all other chapters we have discussed the Israel model. We have sought, empirically, to bring together certain manifestations and to trace their source; they indicate some sort of process which, if fostered and led to its logical conclusion, can serve as the prototype of a complex and many-phased system of vertical integration. We have suggested a method of organization based upon certain combined aspects, underlining the socio-economic trend.

We believe that the advance of technology which has changed the dimensions of the economic unit can stimulate the development of co-operation, to become the shield of the village, protecting it from the encroachments of excessive urban centralization.

A technique can be taught, but the spirit of co-operation must be inculcated. To ensure an adequate rate of progress accordingly, both study and education are necessary.

Our generation lives a very active, intensive and competitive life in technology as in sport, as well as in other useful spheres. Our feeling for social values is becoming weakened. Thus it is incumbent upon us to strengthen the weaker links, to urge Mankind to evolve as Society. The road that must be followed is hard and devious.

In the present volume we have sought to blaze a trail leading perhaps to the main highway.

The principal problem we believe is how to make Mankind more active in the seeking and striving for better ways of life and for better social relations.

We wish to believe that co-operation, fostered through the agency of the *agrindus*, can contribute towards bettering Man's lot on Earth.

203

SOURCES CITED

ABEL Wilhelm. *Agrarpolitik, Göttingen*, Vandenhaeck & Ruprech, 1958, p. 465.

ANDERSON Nels. '*The Urban Community*': *A World Perspective*, Routledge & Kegan Paul, London, 1960, p. 500.

'Changes in the Economic Pattern 1957–9'. The Agric. Register. Oxford University Press, 1960, p. 147.

CHENERY Hollis B. and CLARK Paul G. *Interindustry Economics*, John Wiley & Sons Inc., N.Y., 1959, p. 345.

'City versus County'. *The Economist*, London, 7 Oct. 1961, pp. 18–19.

'Communities in the Making'. *The Economist*, London, 2 Sept. 1961, p. 869.

HIRSH Werner Z. 'Interindustry Relations of a Metropolitan Area'. *The Review of Economics and Statistics*, Nov. 1959, Vol. XLI, No. 4, pp. 360–9.

'Industrialization and rural life in two central Utah Counties', by John R. CRISTIANSEN, Sheridan MAITLAND, John W. PAYNE. Bulletin 416, Utah Exp. Station, Utah State University *et al.*

ISARD Walter, 'Location and Space Economy'. A General Theory relating to Industrial Location, Market Areas, Land Use, Trade and Urban Structure, Massachusetts Institute of Technology, U.S.A., 1956.

ISARD Walter, SHOOLER Eugene W., VICTORISZ Thomas. *Industrial Complex Analysis and Regional Development*. The Massachusetts Institute of Technology, 1959, p. 294.

JOHNSON D. Gale. 'The Dimensions of the Farm Problem'. *Problems and Policies of American Agriculture*, Iowa State University Press, 1960, pp. 47–62.

LÖSCH August. *The Economics of Location*, Yale University Press, 1954.

MOSES Leon H. 'A Genetal Equilibrium Model of Production, Interregional Trade, and Location of Industry'. *The Review of Economics and Statistics*, Harvard University, Nov. 1960, Vol. XLII, No. 4.

'Regional Economics'. John V. KRUTILLE, F. T. MOORE and Discussion. An Economic Review, May 1955, Vol. XLV, No. 2—*Papers and Proceedings*, pp. 120–55.

SCHULTZ Theodore W. 'Value of U.S. Farm Surpluses to Underdeveloped Countries'. *Journal of Farm Economics*, Dec. 1960, Proceedings pp. 1019–30.

SEN S. R. 'The Indian Perspective', ibid., pp. 1031–41.

SOURCES CITED

'Stevenage Growing Up'. *The Economist*, London, 5 Aug. 1961, p. 545.
STEWART H. L. 'The Organization and Structure of Some Representative
  Farms in 1975'. *Journal of Farm Economics*, Proceedings, Dec. 1960,
  pp. 1367–79.
TIEBOUT Charles M. 'Regional and Interregional Input-Output Models:
  An Appraisal'. *The Southern Economic Journal*, Oct. 1957, Vol. XXIV,
  No. 2.
Towards a Capital Intensive Agriculture. U.N. & F.A.O. Parts I and II,
  p. 321, Geneva 1961 (Mimeographed).
'Upper Midwest Economic Study', Sept. 1961. University of Minnesota.
'Voprosi Ekonomiki'. Moscow, No. 10, 1961, pp. 83–91.
Why Labour Leaves the Land, I.L.O., Geneva, 1960, p. 229.

INDEX

INDEX

California, 99, 198
Canada, 6, 10, 123
Cannery, 17, 24, 46, 54, 124, 149, 155, 158, 159; fruit and vegetable, 46, 96, 104
Capital, 39, 100; accumulation of, 193, 194; allocation of, 194; consumption, 191; foundation, 149; goods, 198, 199; human, 198; limited, 149; operative, 23, 114; outside, 67; private, 11, 23-5, 106, 114, 148, 162; share, 23, 24, 75, 96, 149; sources of, 23, 52; structure, 23; working, 24, 36, 56, 125; *see also* operative capital, above
Carmel—Coast, regional council, 72, 80, 167, 168; *agrindus*, 79
Carrots, 162
Cattle, 82, 92; abattoir, 101; breeding, 15, 84, 92, 114, 173, 176; breeders, 173; feed, feeding, feed store, 15, 24, 75, 92; raising, 143; ranch, 92, 93
Chenery Hollis B., and Clark Paul G., 204
China—People's, 8, 42, 198
Citrus, 44, 74, 91, 161, 173; groves, 91, 152, 160, 172, 174; growing, 45, 174-6; packing house, 67, 27, 160; preserves, 159; products factory, 46
Climate, 119, 156, 173, 199
Climatic conditions, 174, 175
Collective, 138, 139; basis, 13; body, 23; character, 182; collaboration, 137; living, 136; methods of production, 27; pool, 20; social organization, 33; system of farming, 196
Collectivism, 138, 180
Collectivistic—initiative, 97; movement, 109, 170; way of life, 40
Collectivization, 198
Commerce, 7, 43, 179; centre of, 1, 145, 146; inspection in, 17
Communal—basis, 32; composition, 145; form of organization, 40; heterogeneity, 2
Commune, 22, 34, 61, 140; autonomous, 42; co-operation between, 186; of communes, 130, 186; people's, 8; regional, 140; urban, 8
Company, 23, 39, 149, 150; collective, 19; construction, 115; co-operative, 19, 44; development, 105, 159; holding, 65, 96, 97; limited, 35, 66, 67, 90, 94, 149; plenary session of, 65, 66; regional economic, 162-3; registered, 54, 64; shares in, 65, 67, 90, 96-99, 162, 163
Consumption, 3, 13, 19, 34, 55, 78, 82, 119, 124, 162, 192; centre, 129; domestic, 29, 44; gross, 191; individualistic, 27; patterns, 194; peak period, 148; private, 43, 191; public, 191, 192, 194
Co-operation, 11, 14, 15, 24, 27, 44, 93, 108, 109, 117, 128, 137, 138, 140, 143, 150, 159, 180, 189, 196-9, 203; consumers', 11; form of, 44, 134; frontiers of, 196; inter-regional, 194; labour, 20; regional, 11, 15, 16, 21, 101, 111, 130, 132, 135-41, 155, 186, 195; spirit of, 159, 203
Co-operative, 16, 19, 21, 22, 66, 140; agriculture, 202; basis, 40, 114, 134, 186, 197; centre, 11, 22; character, 16, 22, 23, 63, 162, 182, 186, 189; consumers', 46; contracting 32; district, 151; enterprise, 12, 44, 82, 104, 111, 139, 188; environment, 22, 180; form, 10, 40, 56; institution, 25; movement, 24, 25; principles, 22, 23, 41, 50, 82, 129; productive, 47; purchase, 13, 27; regional project, 29; 'residents', 21-3; sale, 27; societies, 13, 19-23, 25, 56, 65, 66, 67, 79, 150, 171, 174, 188; store, 20, 23, 144; undertaking, 15, 23, 172
Cotton, 84, 92, 95, 100, 135, 136, 143, 145, 146, 162; crop, 92, 132, 185; cultivation of, 92; gin, 14, 15, 17, 24, 30, 63, 65, 82, 92, 95-7, 100, 101, 104, 133, 134, 174; growing, 99, 174; pickers, 169, 185; raw, 174; seed, 101; spinning mill, 145; spraying, 174
Council—city, 61; county, 61; local, 17, 18, 60, 61, 63, 67-9, 84, 101, 107-11, 114, 128, 130, 167, 168; religious, 69, 70; school, 131

Creamery, 20, 30, 71, 102; regional, 46
Credit, 65, 75; allocation of, 75; conditions, terms, of, 74, 75, 116; security of, 56; volume of, 75
Crop—agricultural, 149; field, 93, 94, 104, 143, 173, 176; industrial, 84, 106, 143; irrigated, 113, 173; rotation of, 114, 173; summer, 83; unirrigated, 91
Cucumbers, 99, 113, 124
Cultivation—area suitable for, 156; intensive, 89, 94, 143; method of, 88, 91; unirrigated, 172
Cultural—activity, 17, 103, 104, 187; centre, 1, 85, 131, 146; character, 192; condition, 28, 195; enterprise, 28, 128; facilities, 15; field, sphere, 15, 138, 140; institution, 120, 128, 131, 138, 153, 184; level, 29, 139, 199; service, 11, 69, 70, 129, 137, 160; standard, 197; undertaking, 108
Culture—Ministry of, *see* Education

Dam(s), 149, 158; *see also* Zohar Dam
Dan—brook, 89
Dates, 15, 114, 117, 173; curing and packing house, 174
Dead Sea, 113
Degania—Aleph, 12, 120, 139; Beth, 125
Deganim Ba'hulah Company, 90, 94, 104
Denmark, 6
Diaspora, 26, 27
Director of the Palestine Office, 12
Dimona, 178, 185
Dorot, 133
Drainage, 18, 63, 65, 84, 86, 88, 89, 93, 106, 168; authority, 127, 172 (*see also* Hulah and Lachish); canals, 109; scheme, 154; system, 114; work, 36, 71, 86, 87, 95

Earners, 43, 44, 106, 162; distribution of, 43; wage, 19, 44, 45
Earnings, 7, 16, 20, 22, 35; equal, 22; scale of, 22
Economic—activity, 15, 65, 66, 192; basis, 108, 143, 182, 201; body, 96; capability, 198, 199; character, 65, 137, 138, 140, 148, 178; condition, 18, 24, 28, 35, 129, 184, 195; consolidation, 182; collaboration, 7, 96, 124, 137; depression, 45; development, 6, 24, 79, 181, 197, 199; enterprise, 18, 19, 20, 23, 24, 28, 46, 63, 65, 67, 71, 125, 128, 137, 185; factor, 148, 190, 193; force, 190; function, 64, 65, 148; institution, 17, 149; level, 13, 42; operation, 11, 18, 64, 66, 193; planning, 18, 97, 151, 190, 192; problem, 63, 66, 191; socio-economic structure, combination etc., 8, 13, 15, 138, 194, 200, 202, 203; stability, 110, 147; structure, 114, 181; ties, 110
Economics—agricultural, 202; of scale, 50, 59, 82, 141, 193, 196
Economist, The, 1, 61, 179, 183, 184
Education, 8, 65, 68-70, 105, 137, 144, 151, 179, 182, 190, 198, 199, 203; adult, 179; Centre, 20; expenditure on, 34, 69 (*see also* expenditure); higher, 104; Ministry of, 18, 184; religious, 151; vocational, 104
Educational—environment, 5; institution, 15, 120, 121, 128, 131, 138, 153, 158, 184; method, 121; principle, 131; service, 20, 144, 160
Egypt, 72, 73, 187
Eilat, 72, 73, 158, 178, 185
Ein Carmel, 80
Ein Gev, 125
Ein Hahoresh, 80
Ein Harod, 33, 35, 38, 40, 54, 170
Ein Shemer, 80
Einat, 80

INDEX

INDEX

Israel Bank (*see also* Bank, Central), 75
Israel Defence Army, 42
Israel Institute of Technology, 104
Israel Water Planning Authority (Tahal), 18, 89, 109, 174

Japan, 199
Javneh, 72, 73, 164, 165, 178, 184
Jerusalem, 26, 36, 37, 39, 45, 62, 72, 73, 85, 142, 151, 165, 166, 176; highlands, mountains, 37, 176; unit, 37, 39
Jewish Agency, 24, 94, 105, 183; Absorption Department, 145; Settlement Department, 85, 103, 109, 144, 158;
Jewish National Fund, 90, 94, 109
Jewish —settlement, 26, 87; settlement company, 87
Jezreel—Valley of, 37, 38, 62, 170, 171, 186; regional council, 72, 73, 168
Johnson D. Gale, 202, 204
Joinery, 45, 47, 57, 102, 146; workers, 125
Joint Agricultural and Settlement Planning Centre, 24
Jordan —*Agrindus*, 129; border, 113, 142, 153; river, 72, 73, 83, 87, 88, 95, 113, 157; Valley, 93, 105, 113–29, 159, 170, 172, 173, 186, 189
Jordan Valley Enterprises Company, 118, 119, 126
Jordan Valley Heavy Equipment (regional station), 115
Jordan Valley packing house, 117
Jordan Valley regional council, 72, 114, 122, 125, 170
Jordan Valley Settlements Organization, 114, 122
Judean —highlands, hills, mountains, 72, 73, 142, 165, 166

Kadesh Naphtali, Vale of, 83
Karkur, 37
Karver, Professor, 147, 148
Kaushan, 197
Kefar Szold, 54
Kelet (Plywood Factory), 54, 122, 123, 124; production, 123
Kfar avoda (labour village), 177
Kfar Gileadi, 36–8, 41
Kfar Glickson, 80
Kfar Hanassi, 54
Kfar Masaryk, 54
Kfar Yehezkel, 170
Kfar Yeruham, 178
Kibbutz, 12, 13, 15, 16, 20, 22, 27, 29, 30, 33, 44, 53, 56, 58, 63, 109, 115, 122, 123–5, 130, 132, 136–40, 155, 170, 180; common pool, 22; co-operation, 137; democracy, 137; enterprises, factory, industry, 29, 30, 54, 124; grouping, 53; inter-*kibbutz* collaboration, 103; left-wing, 40; members, settlers, 5, 16, 20, 22, 27, 38, 55, 56, 85, 105, 124, 137, 139, 140; movement, 33, 74, 104, 109, 136, 137, 150, 170
Kibbutzim, 5, 10, 12, 21, 27–30, 42, 45, 47, 50, 51, 53, 54, 56, 61, 71, 74, 76, 78, 79, 84, 103, 104, 106, 109, 110, 121, 122, 124, 130, 135, 138, 139, 141, 143, 144, 150, 151, 153–5, 168, 172, 176, 180, 186, 187; collaboration of, 9
Kinnarot Lowland, 113
Kinneret—(Sea of Galilee), 112, 113, 125–7; local council, 114; Fishermen's Organization, 125; sub-district, 62
Kirur (Cold and Curing Store), 77, 79
Kiryat Gat, 72, 142, 143, 145, 146–51, 159, 160, 178; planners, 146
Kiryat Melachi, 178
Kiryat Shmona (Halsa), 85, 97, 98, 101, 102, 104–11, 178, 185, 187
Kishon—regional council, 72, 73, 168, 170

Kolkhoz(*y*), 8, 196, 197; artificial insemination station, 197; collaboration, 197; industrial plants, 197
Koor, 45, 46, 102
Kvutza (collective settlement), 12, 53, 139

Labour, 6, 7, 13, 52, 55, 124, 148, 184, 198, 200; agricultural, 6; building, 144; corps (*see also* G'dud Ha'avoda; Council), 184; division of, 197; efficiency, 37, 93, 196, 197; farm, 111, 185, 202; fluctuation, 185; force, 3, 6, 30, 37, 45, 48, 49, 55, 106, 120, 134, 147, 148, 197, 199; groups, 32; hired, 12, 13, 20, 21, 29, 52, 55, 56, 123, 157, 158; manual, 27, 106, 118; movement, 21, 104; organization, 44; outside, 21; productivity of, 15, 19; self, 13, 20, 21, 27, 29, 56, 178; stability, 148
Lachish —brook, 142; Brook Drainage Authority, 142; Development Authority, 146, 149; district, 142–151, 158, 159, 160; regional council, 72, 73, 144, 145, 149, 150, 151; Settlement Department, 146, 150
Lahmenu (Our Bread), 101, 104
Lebanon, 72, 73, 112, 169
Lehavot Habashan, 102
Lehavot Haviva, 80
Lemon, 162
Licensing Authority, 115
Linear programming, 192, 193, 194
Livestock, crops of, 52; feed, 91, 119; feeding operations, 201; health of, 77; increase in, 52; products of, 44
Loan(s), 25, 68, 82, 154; Government, 24, 105; repayment of, and consolidation, 67, 68, 168; security of, 56
Location - advantages of, 129; Economics of, 129, 196; of Economic Activity, 129; Industrial, 129; problem, 193; Theory, 129, 193
Lösch August, 129, 193, 195, 202, 204
Louisiana, 12
Lucerne, 76, 92, 113, 131, 132, 173; cultivation of, 133; dehydrating plant, 96; drying, 102, 103; field, 76, 133; grinding, 76; meal, 76, 133, 173; mills, 14, 29; 76–9, 96, 104, 133, 136, 174; planter, 174
Lydda Plain region, 72, 168

Maabara, Maabarot (transitional work camp(s), 21, 85, 177, 178
Ma'abarot (*Kibbutz*), 80
Ma'agan Michael, 80
Ma'aleh Hagalil, 72, 168
Ma'alot, 178
Ma'anit, 54, 80
Ma'yan Zvi, 80
Machinery, 1, 39, 48, 54, 76, 86, 93, 102, 123, 139; agricultural, 54, 58, 116, 151, 169, 170, 174; farm, 90; heavy, 16, 104
Machines, 3, 11, 39, 44, 47, 98, 101, 123, 174
Magal, 80
Malcha springs, 83, 104
Manpetat Hazafon Ltd. (Northern Cotton Gins), 99–101, 104
Manpower, 15, 19, 23, 43, 95, 107, 135; problem, 50; shortage, 29, 135; wastage of, 201
Ma'on—Hevel (region), 72, 73, 141, 157, 169; regional council, 141, 156
Marble—coloured, 39; deposit of, 38; Italian, 39; quarry, 38
Marit, 130
Market, 38, 100, 106, 117, 119, 147, 166, 172, 173, 193; abroad, foreign, outside, 58, 116, 123, 133; domestic, local, 14, 39, 52, 116, 133, 161; gluts, 29; labour, 35, 56, 85; overseas, 44; national prices, 191

211

INDEX

feed, 14, 15, 24, 75, 77; live, 100, 172; slaughterhouse, 119, ,120 (*see also* House, slaughter); table, 119, 172
Potatoes, 15, 74, 92, 132, 136, 173; autumn, 92; winter, 161, 162
Pri Hagalil Ltd., 99, 101, 104
Produce—agricultural, 19, 71, 106, 143, 158, 190, 194; farm, 15, 17, 38, 51, 59, 66, 96, 99, 102, 114, 143, 154, 160
Product—national, 30; net national, 191
Production, 3, 13, 34, 43, 48, 76, 82, 99, 100, 101, 119, 123, 124, 133, 146, 161, 162, 188, 190, 191, 196, 199; accounting of, 33; agricultural, 13, 44, 122, 128, 129, 138, 147, 174, 196, 197; annual, 74; chemicalization of, 196; collective methods of, 27; cost of, 192, 193, 196; council, 111; electrification of, 196; farm, 15, 29; means of, 76, 161, 191, 192, 198; mechanization of, 196; primary, 190; programme, 93; secondary, 190; value of, 59
Propensity to consume, 194
Purchasing organization, 96, 104

Quarry, 36, 39, 45, 47, 48, 50, 57, 58, 59, 86, 101, 115, 118; marble, 38 (*see also* Marble); stone, 118; *see also* Even Vasid

Rainfall, 83, 110, 156
Ramat Hakovesh, 80
Ramat Rahel, 37, 39, 41
Ramim Works, 102
Ramleh, 62
Region—agricultural, 66; co-operative agro-industrial, 30; economic, 129, 194; highland, 83; united, 151
Regional—administration, 17; associations, 31, 63; authority, 126; bank, 16; basis, 14, 18, 140, 197; body, 14; centre, 166, 185–7; character, 29, 51; clinic, 17; contracting agency, 21; co-operative, 140; council—see below, Regional council; Development Officer, 93; enclave, 111; economic company, 124, 162, 163; Executive, 17, 18; feed mixing plant (*see* Plant; functions), 103, 143; gatherings, 188; hospital, 17; industrial development, 51; industrial enterprise, 51, 176; industry, 56; institution, 21, 153; leader, 130, 132; nursery, 86, 90, 91; organization, 14, 128, 154; planning, 144, 158; plants, 12, 15 (*see also* Plant); project, 12, 23, 24, 29, 82, 103
Regional council, 12, 17–19, 47, 50, 60–9, 71, 74, 78, 79, 82, 84, 85, 89, 101, 103, 111, 130, 135, 145, 150–3, 164, 166–8, 170, 171, 176; chairmen of, 12, 64, 66, 67, 125, 131, 135, 164; executive of, 67, 90, 93; plenary session of, 67
Registrar—of Companies, 64; of Co-operative Societies, 64
Rehovot, 62, 147
Renatia, 79
Returns to scale, 192, 193
Revenue, 67, 70, 115
Revivim, 158
Rochdale, 11
Rosh Ha'ayin—unit, 37
Rosh Hanikra, 117
Rosh Pina, 85, 107
Rothschild, Baron Edmond, 26, 28
Ruhama, 130, 133
Rural—area, 8, 11, 23, 30, 52, 186; atmosphere, 61; centre, 143–6; 149, 150, 158, 160; community, 4, 51; county, 61; district, 23, 43, 109; inhabitant, 19; living, 2; organization, 24; people, 52; population, 3, 25, 199; setting, 7, 125, 180; settlement, 11, 19, 177; slaughterhouse, 63; society, 2, 5, 7, 8, 22, 23; standard of living, 8; urban settlement, 198

Russia, 8, 27, 28, 41, 42, 116, 196, 198

Safad, 30, 83, 105, 112, 178, 184; sub-district, 62; town council, 108
Samaria, 12, 74, 152; *see also* Hefer Valley
School, 22, 85, 103, 121, 145, 188, 196, 197; agricultural (*see* Mikveh Israel Agricultural School); council, 131; evening, 131, 141; farm, 43, 121; high, 121, 144, 182; primary, 131, 144, 182; regional, 15, 168; religious, non-religious, 151; secondary, 131; technical, 55; trades, 121, 188; vocational, 105
Schultz Th. W., 199, 204
Sde Boker, 158
Sde Nahum, 54
Sdot Yam, 80
Sderot—town, 130, 134, 178, 187; local council, 130, 187
Sefen, 122, 124, 128; products, 124
Sen S. R., 205
Service(s), 11, 14, 16, 17, 19, 20, 23, 24, 37, 43, 52, 63–70, 76, 79, 85, 86, 93, 104, 108, 120, 128, 129, 132, 135, 137, 139, 142–5, 149, 154, 159, 160, 162, 164, 185, 187, 188, 191, 192; agricultural, 12, 14, 69, 70, 93; centre, 160; commercial, 147; concentration of, 158; cultural, 11, 129, 137, 160; domestic, 35, 37; economic, 10, 118, 120, 137, 143, 144; enterprise, 24, 115, 162; health, 19, 20, 198; public, 21, 43; religious, 69–71, 105; state, 68–71; technical, 10; veterinary, 103, 107
Settlement—agricultural, 10, 14, 21, 29, 32, 34, 38, 45, 57, 59, 67, 74, 79, 84, 85, 105, 123, 129, 156, 159, 160, 161, 163, 185, 188, 190, 193; associated, 74, 75; authority, 161, 162, 176, 178; collective, 12, 16, 21, 36, 38, 41, 140, 196; co-operative, 21; highland, 83, 89, 175; institution, 86, 88, 134; movement, 17, 178
Sewerage, 85, 131; problems, 154; schemes, 157; regional, 18
Sha'ar Hanegev—region, 130–41, 150, 158, 159, 169, 187; regional council, 63, 72, 73, 131–6, 141, 187; Regional School, 130, 131, 134, 136
Shafir, 72, 150, 151
Sharon, 62, 73, 152, 153, 167; Coast regional council, 72, 79, 166; Central regional council, 72, 79, 80, 152, 166; Hadar, 72, 166; North, 72, 153, 166; South, 152
Sheep—breeding, 176; raising, 83, 84, 114, 143
Shefa'im, 80
Shelahim, 72, 141
Shikun Housing Corporation, 109
Ship to Village scheme, 143, 177
Shlomi, 178
Social—absorption, 182; basis, 201; character, 24, 151, 160, 178; condition, 18, 24, 45, 129, 168, 182, 191, 195; enclave, 130; enterprise, 28; environment, 182; factor, 148, 190, 191, 193; function, 148, 182; integration, 180; organization, 28, 33, 160, 188; planning, 20; principle, 13, 34, 50, 137, 159; problem, 29, 56, 161, 201; reform, 195, 201; relationship, 181, 184, 203; service, 143, 144; solutions, 201; sphere, 14, 15, 137, 140; stability, 144; ties, 110, 143; welfare, 68–71, 144, 145; undertaking, 79, 108
Soil, 3, 4, 51, 83, 84, 88, 92, 113, 143, 175, 199; alluvial, 152; brown, 152, 172; calcareous, 113, 172, 175; chalk, 172; clay, 84, 152; conservation of, 86, 88, 106; cultivation of, 4, 20, 26, 27, 156; deterioration of, 113; erosion, 63, 88; fertile, 24, 110, 153; group, 156; improvement of, 89, 114, 173; kurkar, 136; loess, 156; ownership of, 13; quality, 129; research laboratory, 86; salination of, 89, 114, 173; unirrigated, 173
Sollel Boneh Contracting Corporation, 45, 102, 115
Sovkhoz(y), 196, 197

213